BROOKLANDS BOOKS

MESSERSCHMITT
Gold Portfolio
1954-1964

Compiled by
R.M.Clarke

ISBN 1 85520 2417

BROOKLANDS BOOKS LTD.
P.O. BOX 146, COBHAM,
SURREY, KT11 1LG. UK

Printed in Hong Kong

BROOKLANDS BOOKS

BROOKLANDS ROAD TEST SERIES

Abarth Gold Portfolio 1950-1971
AC Ace & Aceca 1953-1983
Alfa Romeo Giulietta Gold Portfolio 1954-1965
Alfa Romeo Giulia Berlinas 1962-1976
Alfa Romeo Giulia Coupés 1963-1976
Alfa Romeo Giulia Coupés Gold P. 1963-1976
Alfa Romeo Spider 1966-1990
Alfa Romeo Spider Gold Portfolio 1966-1991
Alfa Romeo Alfasud 1972-1984
Alfa Romeo Alfetta Gold Portfolio 1972-1987
Alfa Romeo Alfetta GTV6 1980-1987
Allard Gold Portfolio 1937-1959
Alvis Gold Portfolio 1919-1967
Armstrong Siddeley Gold Portfolio 1945-1960
Aston Martin Gold Portfolio 1972-1985
Austin Seven 1922-1982
Austin A30 & A35 1951-1962
Austin Healey 100 & 100/6 Gold P. 1952-1959
Austin Healey 3000 Gold Portfolio 1959-1967
Austin Healey Sprite 1958-1971
BMW Six Cyl. Coupés 1969-1975
BMW 1600 Collection No.1 1966-1981
BMW 2002 Gold Portfolio1968-1976
BMW 316, 318, 320 (4 cyl.) Gold P. 1975-1990
BMW 320, 323, 325 (6 cyl.) Gold P .1977-1990
BMW 5 Series Gold Portfolio1981-1987
BMW M Series Performance Portfolio1976-1993
Bristol Cars Gold Portfolio 1946-1992
Buick Automobiles 1947-1960
Buick Muscle Cars 1965-1970
Cadillac Automobiles 1949-1959
Cadillac Automobiles 1960-1969
Chevrolet 1955-1957
Chevrolet Impala & SS 1958-1971
Chevrolet Corvair 1959-1969
Chevy El Camino & SS 1959-1987
Chevy II Nova & SS 1962-1973
Chevelle & SS Muscle Portfolio 1964-1972
Chevrolet Muscle Cars 1966-1971
Chevy Blazer 1969-1981
Chevrolet Corvette Gold Portfolio 1953-1962
Chevrolet Corvette Sting Ray Gold P. 1963-1967
Chevrolet Corvette Gold Portfolio 1968-1977
High Performance Corvettes 1983-1989
Camaro Muscle Portfolio 1967-1973
Chevrolet Camaro Z28 & SS 1966-1973
Chevrolet Camaro & Z28 1973-1981
High Performance Camaros 1982-1988
Chrysler 300 Gold Portfolio 1955-1970
Chrysler Valiant 1960-1962
Citroen Traction Avant Gold Portfolio 1934-1957
Citroen 2CV Gold Portfolio 1948-1989
Citroen DS & ID 1955-1975
Citroen DS & ID Gold Portfolio 1955-1975
Citroen SM 1970-1975
Cobras & Replicas 1962-1983
Shelby Cobra Gold Portfolio 1962-1969
Cobras & Cobra Replicas Gold P. 1962-1989
Cunningham Automobiles 1951-1955
Daimler SP250 Sports & V-8 250 Saloon Gold Portfolio 1959-1969
Datsun Roadsters 1962-1971
Datsun 240Z 1970-1973
Datsun 280Z & ZX 1975-1983
The De Lorean 1977-1993
De Tomaso Collection No. 1 1962-1981
Dodge Charger 1966-1974
Dodge Muscle Cars 1967-1970
Dodge Viper on the Road
Excalibur Collection No. 1 1952-1981
Facel Vega 1954-1964
Ferrari Cars 1946-1956
Ferrari Collection No. 1 1960-1970
Ferrari Dino 1965-1974
Ferrari Dino 308 1974-1979
Ferrari 308 & Mondial 1980-1984
Motor & T&CC Ferrari 1966-1976
Motor & T&CC Ferrari 1976-1984
Fiat 500 Gold Portfolio 1936-1972
Fiat Pininfarina 124 & 2000 Spider 1968-1985
Fiat-Bertone X1/9 1973-1988
Ford Consul, Zephyr, Zodiac Mk.I & II 1950-1962
Ford Zephyr, Zodiac, Executive, Mk.III & Mk.IV 1962-1971
Ford Cortina 1600E & GT 1967-1970
High Performance Capris Gold P. 1969-1987
Capri Muscle Portfolio 1974-1987
High Performance Fiestas 1979-1991
High Performance Escorts Mk.I 1968-1974
High Performance Escorts Mk.II 1975-1980
High Performance Escorts 1980-1985
High Performance Escorts 1985-1990
High Performance Sierras & Merkurs Gold Portfolio 1983-1990
Ford Automobiles 1949-1959
Ford Fairlane 1955-1970
Ford Ranchero 1957-1959
Thunderbird 1955-1957
Thunderbird 1958-1963
Thunderbird 1964-1976
Ford Falcon 1960-1970
Ford GT40 Gold Portfolio 1964-1987
Ford Bronco 1966-1977
Ford Bronco 1978-1988
Holden 1948-1962
Honda CRX 1983-1987
Hudson & Railton 1936-1940
Isetta 1953-1964
ISO & Bizzarrini Gold Portfolio 1962-1974
Jaguar and SS Gold Portfolio 1931-1951
Jaguar XK120, 140, 150 Gold P. 1948-1960
Jaguar Mk.VII, VIII, IX, X, 420 Gold P.1950-1970
Jaguar 1957-1961
Jaguar Mk.2 1959-1969
Jaguar Cars 1961-1964
Jaguar E-Type Gold Portfolio 1961-1971
Jaguar E-Type 1966-1971
Jaguar E-Type V-12 1971-1975
Jaguar XJ12, XJ5.3, V12 Gold P. 1972-1990
Jaguar XJ6 Series II 1973-1979
Jaguar XJ6 Series III 1979-1986
Jaguar XJS Gold Portfolio 1975-1990
Jeep CJ5 & CJ6 1960-1976
Jeep CJ5 & CJ7 1976-1986
Jensen Cars 1946-1967
Jensen Cars 1967-1979
Jensen Interceptor Gold Portfolio 1966-1986
Jensen Healey 1972-1976
Lagonda Gold Portfolio 1919-1964
Lamborghini Cars 1964-1970
Lamborghini Countach & Urraco 1974-1980
Lamborghini Countach & Jalpa 1980-1985
Lancia Beta Gold Portfolio 1972-1984
Lancia Fulvia Gold Portfolio 1963-1976
Lancia Stratos 1972-1985
Land Rover Series I 1948-1958
Land Rover Series II & IIa 1958-1971
Land Rover Series III 1971-1985
Land Rover 90 & 110 1983-1989
Lincoln Gold Portfolio 1949-1960
Lincoln Continental 1961-1969
Lincoln Continental 1969-1976
Lotus & Caterham Seven Gold P. 1957-1993
Lotus Sports Racers Gold Portfolio 1953-1965
Lotus Elite 1957-1964
Lotus Elite & Eclat 1974-1982
Lotus Elan Gold Portfolio 1962-1974
Lotus Elan Collection No. 2 1963-1972
Lotus Cortina Gold Portfolio 1963-1970
Lotus Europa Gold Portfolio 1966-1975
Lotus Turbo Esprit 1980-1986
Motor & T&CC on Lotus 1979-1983
Marcos Cars 1960-1988
Maserati 1965-1970
Maserati 1970-1975
Mazda RX-7 Collection No. 1 1978-1981
Mercedes Benz Cars 1949-1954
Mercedes Benz Competition Cars 1950-1957
Mercedes Benz Cars 1954-1957
Mercedes Benz Cars 1957-1961
Mercedes 190 & 300 SL 1954-1963
Mercedes 230/250/280SL 1963-1971
Mercedes Benz SLs & SLCs Gold P. 1971-1989
Mercedes S & 600 1965-1972
Mercedes S Class 1972-1979
Mercury Muscle Cars 1966-1971
Messerschmitt Gold Portfolio1954-1964
Metropolitan 1954-1962
MG Gold Portfolio 1929-1939
MG TC 1945-1949
MG TD 1949-1953
MG TF 1953-1955
MGA & Twin Cam Gold Portfolio 1955-1962
MG Midget Gold Portfolio1961-1979
MGB Roadsters 1962-1980
MGB MGC & V8 Gold Portfolio 1962-1980
MGB GT 1965-1980
Mini Cooper Gold Portfolio 1961-1971
Mini Muscle Cars 1961-1979
Mini Moke Gold Portfolio1964-1994
Mopar Muscle Cars 1964-1967
Morgan Three-Wheeler Gold Portfolio 1910-1952
Morgan Plus 4 & Four 4 Gold P. 1936-1967
Morgan Cars 1960-1970
Morgan Cars Gold Portfolio 1968-1989
Morris Minor Collection No. 1 1948-1980
Shelby Mustang Muscle Portfolio 1965-1970
High Performance Mustang IIs 1974-1978
High Performance Mustangs 1982-1988
Oldsmobile Automobiles 1955-1963
Oldsmobile Cutlass & 4-4-2 1964-1972
Oldsmobile Muscle Cars 1964-1971
Oldsmobile Toronado 1966-1978
Opel GT 1968-1973
Packard Gold Portfolio 1946-1958
Pantera Gold Portfolio 1970-1989
Panther Gold Portfolio 1972-1990
Plymouth Barracuda 1964-1974
Plymouth Muscle Cars 1966-1971
Pontiac Tempest & GTO 1961-1965
Pontiac Muscle Cars 1966-1972
Pontiac Firebird & Trans-Am 1973-1981
High Performance Firebirds 1982-1988
Pontiac Fiero 1984-1988
Porsche 356 1952-1965
Porsche 911 1965-1969
Porsche 911 1970-1972
Porsche 911 1973-1977
Porsche 911 Carrera 1973-1977
Porsche 911 Turbo 1975-1984
Porsche 911 SC 1978-1983
Porsche 914 Collection No. 1 1969-1983
Porsche 914 Gold Portfolio 1969-1976
Porsche 924 Gold Portfolio 1975-1988
Porsche 928 1977-1989
Porsche 944 1981-1985
Range Rover Gold Portfolio 1970-1992
Reliant Scimitar 1964-1986
Riley Gold Portfolio 1924-1939
Riley 1.5 & 2.5 Litre Gold Portfolio 1945-1955
Rolls Royce Silver Cloud & Bentley 'S' Series Gold Portfolio 1955-1965
Rolls Royce Silver Shadow Gold P. 1965-1980
Rover P4 1949-1959
Rover P4 1955-1964
Rover 3 & 3.5 Litre Gold Portfolio 1958-1973
Rover 2000 & 2200 1963-1977
Rover 3500 1968-1977
Rover 3500 & Vitesse 1976-1986
Saab Sonett Collection No.1 1966-1974
Saab Turbo 1976-1983
Studebaker Gold Portfolio 1947-1966
Studebaker Hawks & Larks 1956-1963
Avanti 1962-1990
Sunbeam Tiger & Alpine Gold P. 1959-1967
Toyota MR2 1984-1988
Toyota Land Cruiser 1956-1984
Triumph TR2 & TR3 1952-1960
Triumph TR4, TR5, TR250 1961-1968
Triumph TR6 Gold Portfolio 1969-1976
Triumph TR7 & TR8 Gold Portfolio 1975-1982
Triumph Herald 1959-1971
Triumph Vitesse 1962-1971
Triumph Spitfire Gold Portfolio 1962-1980
Triumph 2000, 2.5, 2500 1963-1977
Triumph GT6 1966-1974
Triumph Stag 1970-1980
TVR Gold Portfolio 1959-1990
VW Beetle Gold Portfolio1935-1967
VW Beetle Gold Portfolio1968-1991
VW Beetle Collection No.1 1970-1982
VW Karmann Ghia 1955-1982
VW Bus, Camper, Van 1954-1967
VW Bus, Camper, Van 1968-1979
VW Bus, Camper, Van 1979-1989
VW Scirocco 1974-1981
VW Golf GTI 1976-1986
Volvo PV444 & PV544 1945-1965
Volvo Amazon-120 Gold Portfolio 1956-1970
Volvo 1800 Gold Portfolio 1960-1973

BROOKLANDS ROAD & TRACK SERIES

Road & Track on Alfa Romeo 1949-1963
Road & Track on Alfa Romeo 1964-1970
Road & Track on Alfa Romeo 1971-1976
Road & Track on Alfa Romeo 1977-1989
Road & Track on Aston Martin 1962-1990
R & T on Auburn Cord and Duesenburg 1952-84
Road & Track on Audi & Auto Union 1952-1980
Road & Track on Audi & Auto Union 1980-1986
Road & Track on Austin Healey 1953-1970
Road & Track on BMW Cars 1966-1974
Road & Track on BMW Cars 1975-1978
Road & Track on BMW Cars 1979-1983
R & T on Cobra, Shelby & Ford GT40 1962-1992
Road & Track on Corvette 1953-1967
Road & Track on Corvette 1968-1982
Road & Track on Corvette 1982-1986
Road & Track on Corvette 1986-1990
Road & Track on Datsun Z 1970-1983
Road & Track on Ferrari 1975-1981
Road & Track on Ferrari 1981-1984
Road & Track on Ferrari 1984-1988
Road & Track on Fiat Sports Cars 1968-1987
Road & Track on Jaguar 1950-1960
Road & Track on Jaguar 1961-1968
Road & Track on Jaguar 1968-1974
Road & Track on Jaguar 1974-1982
Road & Track on Jaguar 1983-1989
Road & Track on Lamborghini 1964-1985
Road & Track on Lotus 1972-1981
Road & Track on Maserati 1952-1974
Road & Track on Maserati 1975-1983
R & T on Mazda RX7 & MX5 Miata 1986-1991
Road & Track on Mercedes 1952-1962
Road & Track on Mercedes 1963-1970
Road & Track on Mercedes 1971-1979
Road & Track on Mercedes 1980-1987
Road & Track on MG Sports Cars 1949-1961
Road & Track on MG Sports Cars 1962-1980
Road & Track on Mustang 1964-1977
R & T on Nissan 300-ZX & Turbo 1984-1989
Road & Track on Peugeot 1955-1986
Road & Track on Pontiac 1960-1983
Road & Track on Porsche 1951-1967
Road & Track on Porsche 1968-1971
Road & Track on Porsche 1972-1975
Road & Track on Porsche 1975-1978
Road & Track on Porsche 1979-1982
Road & Track on Porsche 1982-1985
Road & Track on Porsche 1985-1988
R & T on Rolls Royce & Bentley 1950-1965
R & T on Rolls Royce & Bentley 1966-1984
Road & Track on Saab 1972-1992
R & T on Toyota Sports & GT Cars 1966-1984
R & T on Triumph Sports Cars 1953-1967
R & T on Triumph Sports Cars 1967-1974
R & T on Triumph Sports Cars 1974-1982
Road & Track on Volkswagen 1951-1968
Road & Track on Volkswagen 1968-1978
Road & Track on Volkswagen 1978-1985
Road & Track on Volvo 1957-1974
Road & Track on Volvo 1975-1985
R&T - Henry Manney at Large & Abroad

BROOKLANDS CAR AND DRIVER SERIES

Car and Driver on BMW 1955-1977
Car and Driver on BMW 1977-1985
C and D on Cobra, Shelby & Ford GT40 1963-84
Car and Driver on Corvette 1956-1967
Car and Driver on Corvette 1968-1977
Car and Driver on Corvette 1978-1982
Car and Driver on Corvette 1983-1988
C and D on Datsun Z 1600 & 2000 1966-1984
Car and Driver on Ferrari 1955-1962
Car and Driver on Ferrari 1963-1975
Car and Driver on Ferrari 1976-1983
Car and Driver on Mopar 1956-1967
Car and Driver on Mopar 1968-1975
Car and Driver on Mustang 1964-1972
Car and Driver on Pontiac 1961-1975
Car and Driver on Porsche 1955-1962
Car and Driver on Porsche 1963-1970
Car and Driver on Porsche 1970-1976
Car and Driver on Porsche 1977-1981
Car and Driver on Porsche 1982-1986
Car and Driver on Saab 1956-1985
Car and Driver on Volvo 1955-1986

BROOKLANDS PRACTICAL CLASSICS SERIES

PC on Austin A40 Restoration
PC on Land Rover Restoration
PC on Metalworking in Restoration
PC on Midget/Sprite Restoration
PC on Mini Cooper Restoration
PC on MGB Restoration
PC on Morris Minor Restoration
PC on Sunbeam Rapier Restoration
PC on Triumph Herald/Vitesse
PC on Spitfire Restoration
PC on Beetle Restoration
PC on 1930s Car Restoration

BROOKLANDS HOT ROD 'MUSCLECAR & HI-PO ENGINES' SERIES

Chevy 265 & 283
Chevy 302 & 327
Chevy 348 & 409
Chevy 350 & 400
Chevy 396 & 427
Chevy 454 thru 512
Chrysler Hemi
Chrysler 273, 318, 340 & 360
Chrysler 361, 383, 400, 413, 426, 440
Ford 289, 302, Boss 302 & 351W
Ford 351C & Boss 351
Ford Big Block

BROOKLANDS RESTORATION SERIES

Auto Restoration Tips & Techniques
Basic Bodywork Tips & Techniques
Basic Painting Tips & Techniques
Camaro Restoration Tips & Techniques
Chevrolet High Performance Tips & Techniques
Chevy Engine Swapping Tips & Techniques
Chevy-GMC Pickup Repair
Chrysler Engine Swapping Tips & Techniques
Custom Painting Tips & Techniques
Engine Swapping Tips & Techniques
Ford Pickup Repair
How to Build a Street Rod
Land Rover Restoration Tips & Techniques
MG 'T' Series Restoration Guide
Mustang Restoration Tips & Techniques
Performance Tuning - Chevrolets of the '60's
Performance Tuning - Pontiacs of the '60's

BROOKLANDS MILITARY VEHICLES SERIES

Allied Military Vehicles No.1 1942-1945
Allied Military Vehicles No.2 1941-1946
Complete WW2 Military Jeep Manual
Dodge Military Vehicles No.1 1940-1945
Hail To The Jeep
Land Rovers in Military Service
Off Road Jeeps: Civ. & Mil. 1944-1971
US Military Vehicles 1941-1945
US Army Military Vehicles WW2-TM9-2800
VW Kubelwagen Military Portfolio1940-1990
WW2 Jeep Military Portfolio 1941-1945

1714

CONTENTS

BROOKLANDS BOOKS

ACKNOWLEDGEMENTS

This book on Messerschmitt's unique Kabinenrollers has been one of the most challenging projects that we have tackled for many a year, and also one of the most enjoyable for me personally.

It was challenging because when we started we had less than 50 pages that we could draw on from our own archive. This meant that we had to find at least a further 45 pages to complete even a modest book. Our library will normally produce for us twice as many pages as we need for any given subject and our problem then is to select the best stories to give a comprehensive picture of the marque being covered - so we had an unusual situation.

The enjoyment came in tracking down and obtaining the extra material. This brought me into contact with a wonderful bunch of enthusiasts on two continents whose unstinting help made it possible for us to publish this 172 page Gold Portfolio.

Our first contact was with Phil Boothroyd of the U.K. Messerschmitt Owners Club whose enjoyable stories on Fritz Fend and Willy Messerschmitt will be found within and which first appeared in the M.O.C's fine journal Kabinews. We then contacted Ernie Freestone in the US who had previously assisted us in compiling our volume on Isetta and who was a keen Messerschmitt owner. Ernie not only loaned us valuable technical books and club journals from his collection but also arranged to photograph his splendid KR200 which can be seen on the front cover. He then passed me onto Jim Hockenhull in Oregon. Jim is well versed in all things Messerschmitt and produced a comprehensive list of Kabinenroller stories and other literature which enabled us to locate many articles that were in our own collection but which had not been catalogued.

Jim suggested that I contact two proverbial Messerschmitt affectionados, Peter Svilans in Ontario Canada and Marilyn Felling in California. Peter at short notice sent us a comprehensive package of wonderful well photographed material, some of which appears in the book direct from his copies. Marilyn put together a super collection of sales literature covering all the models and generously entrusted me with them. She also loaned us useful technical text written by her husband Carl and which first appeared in their well known magazine 'Bubble Notes'. The story of Marilyn and Carls KR200 can be found in Tim Howley's article from Special Interest Autos on page 136. Finally, I would like to thank David Kayser who turned my attention to bubble cars in general and to Isetta and Messerschmitts in particular and who is responsible for encouraging me in these 'fun' subjects.

Our thanks as always go to the managements of the following journals who understand the needs of enthusiasts and generously allow us to reissue their copyright road tests and other stories in these anthologies: *Autocar, Autocourse and Sporting Motorist, Autosport, Bubble Notes, Car Life, Classic and Sportscar, Classic Car Mechanics, Cycle, Economy Cars, Custom Car, Kabinews, Light Car, Motor, Motor Cycle, Motorcycle Enthusiast, Motor Cycling, Motor Life, Motor Manual, Motorcycle Mechanics, Motor Trend, Popular Classics, Road & Track, Rod Builder & Customiser, Roller Revue, Scooter and Three Wheeler, Special Interest Autos, Sports Cars Illustrated* and *Wheels*.

R M Clarke

MESSERSCHMITT MAKES A MINICAR

Novel Three-Wheeler with Air-Cooled Engine, Four-Speed Gearbox and Tandem Seating Reveals Several Features Based on Aeroplane Practice

A 56-year-old German aeronautical engineer who provided the Luftwaffe with some of its most formidable fighter aircraft during the war, is (reports Grenville Manton) now concentrating his energies on the production of a minicar of novel design: he is Willi Messerschmitt.

Described by the designer himself as a "Kabinenroller" (or cabin scooter), this little vehicle is a three-wheeled pocket-size fixed-head coupé with two seats arranged in tandem. It has an unusual appearance but is nevertheless very practical.

The engine is a Fichtel and Sachs 174-c.c. air-cooled single-cylinder two-stroke. It is installed in the back and drives the rear wheel through a roller chain. The engine is built integrally with a four-speed gearbox, the unit being mounted on a forked extension of the tubular frame. An unusual feature is the three-plate cork-insert clutch which is engaged and disengaged automatically by centrifugal means.

The front wheels are independently mounted and rubber sprung and the steering is direct and novel in that the conventional steering wheel is replaced by a handlebar, somewhat resembling the control column (or "joystick") of an aeroplane. A simple form of brake gear has been adopted with a two-shoe system on all wheels. The brakes are actuated by a pedal and cable linkage; there is, also, a self-locking handbrake control. The engine is controlled by a throttle twist-grip mounted on the handlebar, but a conventional type of lever placed on the right of the driver's seat provides the gear-change.

Consumption : 123 m.p.g.

The engine (62 mm. by 58 mm.) is started by a kick-starter mounted on the floor and operated from the driver's seat. The specification includes a Bosch flywheel magneto and a dynamo. The compression ratio is 6.6 to 1. At peak r.p.m. (5,260) the engine develops $9\frac{1}{2}$ h.p. It is claimed that the Kabinenroller has a fuel consumption of 123 m.p.g., that it cruises at 47 m.p.h. and has a top speed of 50 m.p.h.

The body is constructed of sheet steel, and as one would expect when it is remembered that Messerschmitt has been designing aircraft for more than 30 years, the form is distinctly aerodynamic and is obviously designed with reduction of drag as a primary aim.

The nose is well streamlined and the wings merge into the panels; the headlights are submerged. As the overall width is only 48 in. and the height 47 in. the frontal area is kept low. The windscreen is of safety glass and a seamless deep-drawn Plexiglass hood fits on to the body. The general lines indicate that the designer has

been influenced by his long association with aircraft practice.

With the Plexiglass sliding windows (which can be removed when desired) the body can be completely enclosed. The seats are

. . . And this is the somewhat curious-looking three-wheeler that Messerschmitt is making. Note the arduous conditions under which it has been tested.

upholstered with rubber-impregnated horsehair and foam-rubber pads and the driver's seat is hinged to facilitate entry and egress

for the occupant of the rear seat. The seat itself, by the way, is quickly removable so that additional luggage space is available when necessary. The tail of the body houses a $2\frac{1}{2}$-gallon petrol tank, plus a quart reserve tank and the spare wheel; above these there is a luggage boot. The standard equipment comprises speedometer, windscreen wiper, driving mirror and blinker-light indicators, and facilities are provided for a radio to be installed.

It is claimed that the low centre of gravity and well-planned weight distribution ensure excellent stability under ordinary road conditions. The wheels (4.00 by 8) are interchangeable and have chromium-plated hub caps. The track measures $36\frac{1}{4}$ in. and the wheelbase 80 in. Other dimensions are: length 111 in. and ground clearance $6\frac{1}{4}$ in. When fully laden the weight of the machine is 790 lb.

The Messerschmitt Kabinenroller is now in full production at a factory in Regensburg.

* * *

As a postscript to Grenville Manton's description we would point out that the manufacturers of the Bond are entitled to the credit for popularizing a Minicar built on really practical lines—and thereby revealing the possibilities of designs of this kind. In Germany the principle was quickly grasped, but the tendency was for such cars to be equipped with over-generous bodywork for the size of engine employed. The photograph clearly shows that Messerschmitt realizes the attendant drawback of this combination; equally, it indicates his original approach to the main problem.

The 174 c.c. Messerschmitt Kabinenroller

A Novel Three-wheeler, Seating Two in Tandem
Good Performance and Excellent Weather Protection

The Fichtel and Sachs 174 c.c. fan-cooled two-stroke engine is mounted at the rear. The rubber suspension unit controlling the rear wheel can be seen behind the strut supporting the silencer

WHEN it was first exhibited at the Geneva Show in March last year, the German Messerschmitt Kabinenroller (Cabin Scooter) created something of a furore. Nothing quite like it had ever been seen before, and its aircraft ancestry was apparent in every line. On the Continent, where folk have fewer inhibitions regarding personal transport than we have in Britain, this little three-wheeler has firmly established itself and is now being produced at a rate of over 200 a week.

Arrangements were completed a few months ago to import the Kabinenroller into Great Britain. So unorthodox a vehicle merits a brief description. There is a simple, tubular, basic frame to which is attached the stressed-skin, pressed-steel body. The engine unit is carried on a rubber-mounted sub-frame at the rear; this sub-frame incorporates the rear-wheel-suspension pivots which are on the sprocket axis, so that wheel movement does not result in chain-tension variations.

Tandem seating was chosen to provide a slim, well-streamlined shape and a reasonably narrow track without loss of stability. Since the passenger's legs lie alongside the driver's seat, leg room is ample without excessive body length. The Plexiglass hood embodies a safety-glass front panel and hinges sideways, giving easy entrance and exit; there are two sliding Plexiglass windows on each side of the hood.

When the hood is opened, a small luggage compartment is revealed above the engine. For solo travel, the rear seat can be removed to provide additional luggage space. The rear panel hinges upward, revealing the 174 c.c. Sachs two-stroke engine and fuel-tank with a readily accessible filler.

Both car and motor-cycle type controls are employed. Clutch and brake pedals occupy conventional car positions, and the hand brake is on the right, forward of the driver's seat. Steering is by handlebar, and the twistgrip throttle control is on the left side of the bar. Manually operated, the gear change is of the positive-stop type, and the lever is convenient to the driver's right hand. An electric starter is fitted and produces ample power to turn the engine briskly. Starting was instantaneous whether the engine was hot or cold. The remote tickler for cold starting is operated by a convenient control at the rear of the passenger compartment. The strangler control knob is situated in the dash. As soon as the chill was off the engine it would tick over reliably and unusually slowly.

Once under way, the first impressions were of the solid manner in which the Kabinenroller held the road, and of the extreme accuracy of the steering. In motor-cycle fashion, steering is direct so that movement of the handlebar corresponds to movement of the front wheels. During normal driving, one finds that steering is done by an almost unconscious gesture of the shoulders rather than by arm action, and the vehicle can literally be placed to an inch.

A straight course was held entirely without effort save in gusty cross winds, when the light weight caused a slight tendency to wandering which, however, was easily corrected. A virtue of the low centre of gravity and the suspension characteristics was that corners could safely be taken at speeds which some might consider dangerously high for a sidecar outfit or a car.

The steering was strongly self-centring, and the limiting factor when cornering on dry roads seemed to be the physical effort required to deflect the front wheels, through the low leverage of the narrow handlebar, rather than tyre adhesion or lifting of the inside wheel. On wet roads, deliberate violence induced mild, though readily controllable, front-wheel sliding.

By modern motor-cycle standards, the suspension was definitely hard and resulted in some jolting on inferior surfaces. The

Though its overall dimensions are small, the Messerschmitt seats two average-stature adults comfortably. Excellent all-round visibility is provided by the Plexiglass "cockpit" cover

sprung driver's seat took much of the sting out of the bumps but the passenger had a less comfortable ride. It is understood that the springing is to receive attention.

The 174 c.c. engine is claimed to have a peak output of over 9 b.h.p., and it certainly endows the Kabinenroller with a more-than-adequate performance for an economy vehicle. As would be expected, it was necessary to use the wide-ratio gear box freely to keep up with normal traffic in towns and in hilly country; driving accordingly (and losing no time on corners) no difficulty was found in averaging about 30 m.p.h. under give-and-take conditions, which makes the Messerschmitt suitable for quite serious motoring.

At one end of the scale, bottom gear gave an ample power margin for the final 1 in 5 section of a Surrey hill, while speedometer readings of over 80 k.p.h. (about 50 m.p.h.) were easily obtainable in top gear on the level, without assistance from the wind. Comfortable cruising speed lay anywhere between

Access and egress are gained by means of a hinged, upper-body section. Sliding windows are fitted in the sides of the transparent top

INFORMATION PANEL

ENGINE: Fichtel and Sachs fan-cooled, single-cylinder two-stroke, 174 c.c. (62 x 58mm); aluminium-alloy cylinder head; compression ratio 6.6 to 1; petroil lubrication.

CARBURETTOR: Bing needle-jet type, with air filter and twistgrip throttle control; remote controls for strangler and tickler.

TRANSMISSION: Four-speed gear box in unit with engine, driven by gears through foot-controlled, cork-lined clutch. Gear ratios: Top, 4.8 to 1. Third, 6.2 to 1. Second, 9.3 to 1. Bottom, 16.2 to 1. Positive-stop hand control to gear change, with neutral-finding lever. Final drive by partially enclosed ½ x 0.305in chain.

BRAKES: Independent foot and hand operation on all three wheels; drums approximately 4½in diameter x 1⅛in wide.

SUSPENSION: All wheels sprung by cylindrical bonded-rubber units with air damping. Front wheels carried on transverse pivoted arms; rear wheel mounted on arm formed by chaincase, pivoting about gear-box sprocket centre.

WHEELS: Interchangeable, pressed-steel, stub-axle mounted, fitted with 4.00 x 8in Michelin tyres.

ELECTRICAL EQUIPMENT: 12-volt Siba combined dynamo, starter and cooling-fan unit forming engine flywheel; ignition by battery and coil; automatic voltage control for dynamo. Built-in headlamps with twin-filament 25/25w bulbs and separate parking light bulbs in main reflectors; twin tail lamps and separate stop light; winking direction indicators; electric horn.

FUEL TANK: 11 litres (approximately 2.4 gallons) capacity, with 1½ quarts reserve. Fuel filter embodied in outlet union.

DIMENSIONS: Wheelbase, 80in; track, 36¼in; overall length, 111in; overall width, 48in; height, 47in; ground clearance, 6¼in; weight, including approximately one gallon of fuel, 406lb.

PRICE: Basic, £274 15s; total price, including purchase tax (in Britain only), £335 18s 2d. Delivery charge, £13. Extras: spare wheel, £2 10s; jack, 15s. Annual tax, £5.

MANUFACTURERS: Regensburger Stahl- und Metallbau Gmbh, Messerschmitt-Werk, Regensburg, Germany.

SOLE CONCESSIONAIRES FOR GREAT BRITAIN AND THE COMMONWEALTH: Beulah Hill Engineering Co., Ltd., 411, Beulah Hill, London, S.E.19.

racing-type upward changes sometimes resulted in an audible click from the mechanism. Owing to the rather lengthy, and therefore whippy, linkage to the gear box, bottom gear would not always engage from neutral at the initial attempt. Selection of the neutral between any pair of gears is effected by the actuation of a trigger on the gear lever.

Some criticism must be directed against interior noise, though this is another matter which is receiving attention by the manufacturers. Above about 30 m.p.h. in top gear, and corresponding speeds in the lower ratios, a considerable amount of booming and drumming was audible to the occupants. One became accustomed to this after a short time but it made conversation difficult.

Ample Driving Light

No lengthy after-dark trips were undertaken during the test, but on normal out-of-town running the double-dipping headlamps were found to give ample illumination on both main and dipped beams for the speed of the machine. Winking direction indicators are fitted and, since they are non-cancelling and out of the driver's line of vision, the warning light on the dash proved a useful aid.

When the Kabinenroller was driven or standing for any length of time in heavy rain, slight leakage took place through the hole provided for the windscreen wiper spindle. The wiper, incidentally, is manually operated. The handlebar trigger supplied for the purpose was too short and the return spring too strong. To counter criticism with praise, the ample main and reserve capacities of the fuel tank were greatly appreciated; the reserve enabled over 30 miles to be covered. Fuel consumption averaged about 85 m.p.g. during the period of the test including plenty of town running and fairly hard driving on the open road.

At £335 plus, the British price of this most entertaining and highly original three-wheeler is admittedly high, owing to import duty and purchase tax; however, its handiness and practicability, and low running costs, do much to offset the initial outlay.

30 and 45 m.p.h. though the engine did not object to sustaining a speed higher than 45 m.p.h. Maxima in second and third gears were about 28 and 41 m.p.h. respectively.

Too much stress, however, should not be laid on sheer performance. The Kabinenroller is intended primarily as a runabout, and for such duties as would be accomplished by a normal scooter. With its satisfactory low-speed torque and ability to pull down to 16 m.p.h. in top gear, its good steering lock and small dimensions, it can be induced through traffic with the slickness of four-wheelers of many times its power. The absence of a reverse gear was seldom a handicap, for the low weight made pushing on a level surface a matter of finger pressure only.

Used from low or high speed, the brakes were well up to the standard of the modern car or solo motor cycle. An excellent balance between front and rear brakes has been chosen, because under normal braking there was no tendency for the front wheels to lock before the rear wheel, or vice versa; in emergency stops for test purposes, the rear wheel would lock. Wheel adhesion was so good on wet roads that surprisingly rapid deceleration was possible without skidding. The handbrake held securely on the steepest gradient encountered (1 in 5).

Very quick gear changes up or down were possible, although

Steering of the Kabinenroller is by handlebar. The throttle is operated from a twistgrip on the left of the bar

Sonderdruck
aus der
Roller Revue
Heft 2 / 2. Jhrg.

Kabinenrollerkolonne am Hamburger Hafen

Foto: Pierer

DER MESSERSCHMITT-KABINENROLLER UNTER DER LUPE

Roller Revue-Tester fuhr bei Schnee, Regen und viel Gegenwind den KR 175

So kurz vor dem Heiraten erkundigt man sich meistens ganz vorsichtig nach den eventuellen Vätern und Müttern der Angebeteten. So möchte auch der Käufer eines Rollers ganz gern wissen, woher sein Auserwählter (Roller) kommt. Nun, der Kabinenroller braucht sich seiner Abstammung nicht zu schämen, kommt er doch aus den Messerschmitt-Werken in Regensburg. Das ist beruhigend und erfreulich zugleich. Beruhigend, weil hinter diesem Werk die nötigen Mittel und Möglichkeiten stehen, den einmal beschrittenen Weg bis zur letzten Reife zu gehen. Erfreulich, weil dieses Werk von keinem anderen Fahrzeug her vorbelastet ist, und so unbeschwert seine eigenen Konstruktionstendenzen entwickeln kann. Steht dann noch ein Konstrukteur wie Herr Fend, der nicht nur ein großes Wissen, sondern auch eine ehrliche Begeisterung für seine Schöpfung mitbringt, hinter dem Ganzen, so kann man beruhigt in die Zukunft schauen.

Mit dem Kabinenroller irgendwo anzuhalten ist gefährlich. Sie sind nämlich sofort von einer Menge Leute umringt, die ihnen ein Loch in den Bauch fragen. Und über diesen neckischen Fragespielen versäumen sie jeden Termin und jedes Rendezvous. Aber zwei Dinge hat dieser unfreiwillige Galluptest doch klar zutage gebracht:

1. Der Messerschmitt-Roller liegt goldrichtig in seiner Konstruktion, um die Lücke auszufüllen, die zwischen Zweirad und Auto nun einmal zwangsläufig gegeben ist.
2. Es herrschen völlig falsche Begriffe darüber, was der Kabinenroller eigentlich sein will und auch nur sein kann.

Diese Begriffe auf den richtigen Nenner zu bringen und klarzustellen, was von dem Kabinenroller überhaupt erwartet werden kann, ist so wichtig, daß es hier vor Beginn des eigentlichen Testes ausführlich behandelt werden muß. Ein Satz, den ich immer wieder von den schaulustigen Interessenten hörte, war nach einer mehr oder weniger sorgfältigen Betrachtung: „Schade, das ist ja gar kein richtiges Auto!" Also, meine Herrschaften, er soll ja auch gar kein Auto sein, nicht einmal ein Kleinstauto. Natürlich ist es auch kein Roller im Sinne des Wortes mehr, aber die Beurteilungsperspektive muß auf jeden Fall vom Zweirad herkommen, niemals vom Auto. So betrachtet, ist der Messerschmitt-Roller das Erfreulichste, was in dieser Richtung je auf die Beine gestellt wurde. Überlegen Sie doch nur: Der Kabinenroller hat alle Vorteile des zweirädrigen Kollegen, die billigen Unterhaltskosten, die einfache Unterstellmöglichkeit, die Fahrerlaubnis mit dem Führerschein 4, und nicht zuletzt die einfache Handhabung im Fahrbetrieb. Zusätzlich wird ihnen geboten:

Der vollständige Schutz gegen Witterungseinflüsse, wobei auch Gepäckstücke eingeschlossen sind, Einbaumöglichkeit einer Heizung — im Winter ein nicht zu unterschätzender Vorteil, die bequeme Fahrweise auf Sitzen mit Lehnen und das Fahren auf drei symmetrisch angeordneten Rädern, welches die Lenkung vereinfacht und den Sicherheitsfaktor im Fahrbetrieb um 100% erhöht. In diesem Zusammenhang möchte ich gleich noch zu dem Preis des Fahrzeugs Stellung nehmen. Auf den ersten Blick scheint dieser mit 2375.— DM ab Werk erheblich über dem Rollerniveau zu liegen. Kalkuliert man aber die beim offenen Roller nötige Wetterbekleidung mit ein, außerdem die verschiedenen Zubehörteile, wie Windschutzscheibe, Rückblickspiegel, Reserveradhülle usw., so schmilzt die Differenz für einen im Gesamten gesehenen, nicht mehr bedeutenden Betrag zusammen. Und nun zum Fahrzeug selbst: Als Träger des Ganzen dient ein geschweißter Stahlrohrrahmen, dessen Steifheit nichts zu wünschen übrig läßt. Die vorderen Räder sind einzeln aufgehängt und durch Gummi abgefedert. Die Federwege können als vollkommen ausreichend bezeichnet werden, auch die Federungshärte ist richtig gewählt worden. Die Lenkung wirkt auf die Vorderräder durch Achsschenkellenkung. Am Anfang wirkt die direkte Lenkungsweise etwas ungewohnt, später ist sie nicht weiter störend. Durch die direkte Übersetzung bestimmt, muß sie relativ hart sein. Als Räder werden Rollerräder in der Größenordnung 4 × 8 Zoll verwendet. Die Karosserie darf in ihrer Formgebung als geglückt bezeichnet werden. Der Platz ist für den Fahrer, auch wenn dieser mit einer sehr großen Figur behaftet ist, reichlich bemessen, auch der Beifahrer kann bequem sitzen. Nur der vorgesehene Raum für die Beine beiderseits des vorderen Fahrersitzes ist etwas knapp geraten. Eine Verbreiterung der Karosserie ist jedoch nicht zu empfehlen. Ein Ding mit Pfiff ist die Gepäckunterbringung im Innenraum. Der vorgesehene Gepäckraum nimmt gerade zwei gefüllte Aktentaschen auf, neben dem Beifahrer ist noch Platz für einen kleinen Koffer. Und im Fahrerraum findet sich ebenfalls noch Gelegenheit, kleinere Gepäckstücke unterzubringen. Wie sie also sehen, läßt sich ziemlich viel Gepäck unterbringen, nur muß es in kleine Einheiten aufgeteilt sein. Große Stücke können außen am Gepäckträger befestigt werden, der gleichzeitig als Skihalter dient. Selbstverständlich ändert sich das grundlegend, wenn der hintere Platz nicht besetzt ist, da dann der Sitz herausgenommen werden kann und damit ein wirklich groß bemessener Gepäckraum zur Verfügung steht. Zu einem kleinen Problem gestaltet sich die Mitnahme eines Kindes. Bei einem 4–6jährigen Kinde mag es ja noch gehen, wenn der eigens dafür vorgesehene verbreiterte Rücksitz verwendet wird, bei älteren Kindern wird die Platzfrage zur Qual. Wohltuend empfindet man die freie Sicht nach allen Seiten durch die Plexiglashaube, auch Verzerrungserscheinungen, wie sie gewölbtes Glas hie und da hervorruft, konnten nicht festgestellt werden. Bleiben nur noch die Bedienungshebel zu erwähnen, die in ihrer Anordnung teilweise nicht ganz der Norm entsprechen, was aber eine reine Gewöhnungssache ist. Etwas unglücklich ist die Lage des Hebels der Handbremse, der während der Fahrt nicht ganz einfach erreicht werden kann.

Bei dem verwendeten Einbaumotor handelt es sich um den 175-ccm-Sachsmotor, der bei 5250 Umdrehungen pro Minute 9 PS Dauerleistung abgibt, diese ist für den Betrieb des Kabinenrollers gerade richtig, ein stärkerer Motor wirft zu große Probleme im Fahrgestellbau auf. Geradezu erstaunlich ist das große Durchzugsvermögen dieses Motors in den unteren Drehzahlbereichen, das nicht nur ein weitgehend schaltungsfreies Fahren ergibt, sondern auch bei Schneeglätte und ähnlichem schlüpfrigem Untergrund das Fahren wesentlich erleichtert. Der Motor ist mit dem Getriebe zu einem Block zusammengebaut. 2., 3. und 4. Gang sind in ihrer Abstufung sehr gut gewählt, der 1. dürfte noch kleiner übersetzt werden. Auch die Kupplung sollte für Extremfälle verstärkt werden. Der Antrieb erfolgt auf das Hinterrad mittels freiliegender Kette.

Nach diesem kurzen Steckbrief komme ich zu den Erfahrungen, die die Prüfung während der Fahrt mit sich brachte. Zuerst möchte ich grundlegend feststellen: Da der Kabinenroller absolutes Neuland darstellt, ist es nicht möglich, Beurteilungen im Vergleich zu anderen Fahrzeugen zu treffen. Man kann hier vom Technischen her Kritik üben, die zwar vom praktischen Fahrbetrieb ausgelöst wird, aber eben nur von theoretischer Überlegung belegt werden kann. Die hier ausgeübte Kritik soll also keine absolute sein, sondern hauptsächlich der Anregung dienen, eventuell einen besseren Weg zu finden, dessen Zweckmäßigkeit aber nur die Praxis erhärten kann. Im Grunde genommen ist der Tester in der gleichen Lage wie der Mann, der sich ein neues Fahrzeug kauft. Eines Tages sitzt man auf einem völlig fremden Fahrzeug, mit dem man sich auf Du und Du stellen möchte. (Ich kann verraten, daß es beim Messerschmitt eine ganz dicke Freundschaft wurde.)

Als ich auf dem Fabrikhof den Roller übernehmen sollte, bemühten sich verschiedene Herren im Schweiße ihres Angesichts sehr heftig, das Ding zum Laufen zu bringen, leider vergeblich. Der Me 53 war ausgesprochen dagegen. Ich bekam dumpfe Vorahnungen, denn bei mir mußte er im Freien übernachten und die Nächte waren empfindlich kalt, die Außentemperatur sank bis — 15° ab. Gott sei Dank waren meine Befürchtungen umsonst. Am anderen Morgen kam er nach kräftigem Tupfen auf den 2. Tritt, und so hat er's die ganze Zeit gehalten, die er bei mir war, obwohl er niemals eine Garage gesehen hatte.

Ganz gegen meine Vermutung ging das Einsteigen ohne Schwierigkeiten und Verrenkungen. Später wollte ich es ganz genau wissen und bat einen älteren Mann, der eine Verletzung an der Wirbelsäule hat, doch mal einzusteigen. Dieser Mann muß ein modernes Auto ganz vorsichtig besteigen, um harte Schmerzen zu erleiden. Auch beim Messerschmitt ging es einwandfrei. Ihre Sorgen in dieser Richtung sind also völlig unberechtigt. Ich habe übrigens die Vorrichtung zum Sitzhochklappen beim Einsteigen nie benutzt und halte sie für vollkommen überflüssig. Und nun ging's los mit der Fahrt. Das Drehgas ist links und die Schaltung mit der Kupplung verbunden in der Weise, daß man den Schalthebel zurückzieht und langsam nach vorne losläßt, wobei die Kupplung langsam eingreift und auch der Gang einrastet. Der Leerlauf wird durch einen kleinen Hebel am Lenker eingeschaltet. Das ist am Anfang natürlich ungewohnt, aber wenn

Hier sehen Sie den Roller von Herrn Fend, des Konstrukteurs des KR 175. Beachten Sie bitte den seitlichen Gepäckträger

Der Motor von der Gebläseseite her gesehen. Deutlich zu erkennen der großdimensionierte Schalldämpfer und die freiliegende Kette

Die Radaufhängung bei abgenommenem Kotflügel. Oberhalb, in die Karosserie eingebaut, die Hupe

man's mal raushat, geht's prächtig. Da man die Hand weder vom Gas noch vom Schalthebel zu nehmen braucht, kann man blitzschnell schalten, dafür geht's auch so einfach, daß der Dümmste nichts falsch machen kann.

Als vorsichtiger Mensch, der sein Leben heiß und innig liebt, galt mein nächster Versuch den Bremsen. Sie greifen sehr weich und zügig, ohne dabei an Wirkung zu verlieren. Die Verzögerungswerte liegen weit über den polizeilich vorgeschriebenen Zahlen. Aber Bremsen sind nun mal mein Steckenpferd, und so wünsche ich mir auch hier noch etwas kürzere Bremswege, vor allem bei Gewaltbremsungen aus der Höchstgeschwindigkeit. Der nächste Versuch galt dem Benzinverbrauch (schließlich geht das auf unseren Geldbeutel los). Bei einer Messung über 200 km, vorwiegend im Stadtverkehr gefahren, allerdings bei scharfer Fahrweise, kamen etwas über 3 Liter raus. Bei einer Vollgasfahrt auf der Autobahn, wobei die Fahrverhältnisse durch tiefen Schnee und Durchquerung zahlreicher Schneewehen sehr erschwert wurden, kam ich weit über die Vierlitergrenze. Hier handelt es sich um einen absoluten Spitzenwert nach oben, der auf die Dauer nicht aufrecht erhalten werden kann. Selbst bei schärfster Fahrweise dürfte auf die Dauer die 3½-Litergrenze nicht überschritten werden. Eine der häufigsten Fragen gilt der Gepäckunterbringung. Über die Raumverhältnisse wurde eingangs schon Auskunft gegeben. Die größeren Gepäckstücke müssen, wie gesagt, außen am Gepäckträger angebracht werden. An dem vom Werk als Zubehör gelieferten Gepäckträger habe ich allerdings zweierlei auszusetzen: 1. ist die herausklappbare Auflage zu kurz, das zu ändern wäre eine Kleinigkeit, 2. gefällt mir die seitliche Anbringung nicht. Hinten am Heck wäre er mir lieber. Allerdings müßte dann beim Tanken das Gepäck abgenommen werden. Wenn man aber den Tankeinfüllstutzen durch die Karosserie nach außen führen würde, wäre auch diese Klippe überwunden.

Motor

Hersteller	Fichtel & Sachs
Bauart	Gebläsekühlter Einzylinder-Zweitaktmotor
Spülverfahren	F & S-Gegenstromspülung
Zylinderbohrung	62 mm
Kolbenhub	58 mm
Zylinderinhalt	174 ccm
Verdichtung	6,6:1
Vorzündung	5 mm v. o. T.
Vergaser	Bing-Kolbenschiebervergaser 1/24 mit Naß-luftfilter und Luftklappe. Leerlauf- und Haupt-düse von außen zugänglich.
Vergasereinstellung	Hauptdüse 115, Leerlaufdüse 35, Nadeldüse 1508. Nadel in der 2. Kerbe.

Getriebe, Antrieb

Art	Vierganggetriebe mit Handschaltung
Kupplung	Dreischeiben-Korklamellen-Kupplung
Übersetzungen	Von der Kurbelwelle zur Vorlegewelle 2,12:1. Von der Vorlegewelle zur Hauptwelle 1. Gang 3,22:1, 2. Gang 1,85:1, 3. Gang 1,24:1, 4. Gang 0,95:1
Kraftübertragung	Rollenkette ½ × ⁵/₁₆ auf Hinterrad
Starter	Starter von innen und außen

Fahrgestell

Rahmen	Schweißkonstruktion aus Präzisionsstahlrohr
Lenkung	Vorderräder gelenkt durch Achsschenkel-lenkung. Wendekreis 8 m.
Bereifung	4,00 × 8 (sämtliche Räder gegeneinander aus-wechselbar)
Luftdruck	Vorne: 0,75 atü, hinten 1,5 atü
Federung	Vorne Einzelradaufhängung mit Gummi-Federung
Bremsen	Durch Seilzug betätigte Innenbackentrommel-bremsen. Fuß- und feststellbare Handbremse auf alle drei Räder wirkend

Elektr. Ausrüstung

Zündung	Bosch-Schwungrad-Lichtmagnetzündung 6 V 45 Watt
Scheinwerfer	Durchmesser 105 mm, 15/15 Watt Biluxlampen
Rücklicht	Doppelseitige Nummernschildbeleuchtung mit Schlußlicht und Katzenaugen. Stopplicht
Richtungsanzeiger	Blinklichtanlage mit Kontrollampe
Horn	Elektrisch
Batterie	6 V 6,9 Ah

Karosserie

Haube	Ganzstahlkarosserie Kabinenhaube aus nahtlos gezogenem Plexiglas
Fenster	Große seitliche Schiebefenster aus Plexiglas (herausnehmbar), Frontscheibe aus Sicherheits-glas
Gepäckraum	Abschließbarer Gepäckraum im Heck, ca. 500 mm tief, 600 mm breit und 170 mm hoch.
Sitze	Bequemer, hochschwenkbarer Vordersitz, Begleitersitz herausnehmbar, dadurch zusätz-licher Gepäckraum. Polsterung beider Sitze aus gummiertem Roßhaar mit Schaumgummi-auflage

Maße und Gewichte

	Länge 2820 mm, Breite 1220 mm, Höhe 1200 mm, Radstand 2030 mm, Spurweite 920 mm
Bodenfreiheit	160 mm
Gewicht	175 kg
Tankinhalt	11,5 l Benzin-Ölgemisch 1:25, davon 1,5 l Reserve
Leistung	Dauernd 9 PS bei 5250 U/min, kurz 9,5 PS bei ca. 5250 U/min
Kraftstoff-Normverbrauch	2,3 l/100 km
Geschwindigkeit	Spitze ca. 80 km, autobahnfest 75 km
Serienausrüstung	Tachometer, Scheibenwischer, Rückspiegel, Werkzeug
Sonderzubehör	Reserverad, verchromte Radzierkappen, Zeit-uhr, Pflegebeutel für Plexiglas, Reifenluftdruck-prüfer, Gepäckhalter, Spezialautosuper, Son-nendach, Sonnenblende, **elektr. Anlasser,** Warmluftheizung

---

Innenraumansicht von oben gesehen, hinter dem 2. Sitz befindet sich die Ablage für den Gepäckraum

Leider war es infolge der schlechten Bodenverhältnisse, es lag dauernd Schnee, nicht möglich, die Bergsteigfähigkeit bis zur Grenze auszuprobieren. Immerhin schaffte er trotz Schneeglätte noch eine Steigung von 20%, mit zwei erwachsenen Personen belastet. Weitere Versuche an stärkeren Steigungen wollte ich nicht vornehmen, da ich ein Durchdrehen des Hinterrades befürchtete. Die 28% bei voller Belastung im ersten Gang glaube ich aber ohne mit der Wimper zu zucken. Die erzielten Reisedurchschnitte, auf der Landstraße gemessen, lagen trotz der Schneeglätte um die 60 km. 40 km Autobahnfahrt ergaben ein Mittel von 72 km. Die absolute Höchstgeschwindigkeit wurde mit 82 km gemessen. Die gefahrenen Schnitte konnten aber auch nur auf Grund der einwandfreien Straßenlage des Fahrzeugs erzielt werden. Ich habe absichtlich versucht, den Roller durch Überziehen in den Kurven zum Kippen zu bringen, es ist mir nicht gelungen. Man kann bei glatten Kurven den Lenker voll einschlagen und mit vollem Dampf reingehen, das Hinterteil kommt brav herum und läuft wieder den Vorderrädern nach. Lediglich bei schnellen Halbkurven machte sich ein Abtreiben nach außen bemerkbar. Ich möchte noch erwähnen, daß die gesamten Fahrten mit einer Belastung von zwei Personen durchgeführt wurden.

Überrascht war ich von dem guten Licht, das ich den 2×15-Watt-Birnen niemals zugetraut hätte, aber es leuchtet die Straße seitlich vollkommen aus, auch die Reichweite genügt vollständig.

Mit der jetzigen Lichtmaschine bin ich nicht ganz einverstanden. Ein Gleichstromaggregat hätte doch viele Vorteile. Genau so wie ich eine kleine Beleuchtung des Innenraumes sehr vermißt habe, dürfte auch die Hupe etwas wirksamer sein. Warum man den Einbau eines Radios vorgesehen hat, ist mir vollkommen unverständlich. Abgesehen davon, daß durch den Betrieb eines solchen die elektrische Anlage stark überlastet wird, ist ein genußvolles Abhören desselben sowieso nicht möglich, da die Motorengeräusche im Innern doch zu stark sind. Sicher, die Karosserie dröhnt in keiner Weise und der Motor ist gut gedämpft, aber der Lärm ist doch so stark, daß man ziemlich laut sprechen muß, um sich mit seinem Beifahrer verständigen zu können.

Anders die Heizung, hier kann man wirklich nicht mehr von Luxus sprechen, sie ist eine Notwendigkeit im Winter und sei's auch nur um die Scheibe zu entfrosten, denn das dauernde Wischen macht einen auf die Dauer verrückt. Der Scheibenwischer mit Handbetrieb durch Zughebel am Lenker betätigt ist zwar eine Notlösung, erfüllt aber letzten Endes seinen Zweck. Das ist also der Kabinenroller, wie er leibt und lebt, ein Fahrzeug, das man mit gutem Gewissen kaufen kann. Vielleicht könnte hier und dort noch eine kleine Verbesserung „angebracht werden, aber das soll den guten Gesamteindruck nicht schmälern. Nur eine Bitte noch an das Werk: Baut keine sogenannten Verbesserungen ein, die letzten Endes nur der Mode entspringen und im Endeffekt den Preis höher treiben. Es soll doch kein Auto werden — was wir brauchen, ist der billige Kabinenroller.

(E. v. F.)

Auch bei solchen Witterungsverhältnissen tat der Messerschmitt-Roller das, was man von ihm erwarten konnte

*S*omebody

just opened

a window

on the

future

NAL TRANSPORTATION TODAY!

All-Purpose

Personal

Transportation for

Commuters

Light Delivery

Suburban Shoppers

Touring

Around the Campus

100 MILES
ON A SINGLE
GALLON OF GAS

ALL-WEATHER COMFORT —seamless plexiglas hood, sliding windows, safety glass windshield —lifts for easy entrance. Independent wheel suspension, sealed underbody, generous leg space.

DEPENDABILITY — the Messerschmitt has won 18 trophies, 28 gold, 6 silver, 5 bronze medals in European endurance and distance trials. Thousands are in use all over the world today.

45-50 MPH CRUISING SPEED —air cooled, rear mounted engine delivers up to 60 mph, swallows hills. Electric starter, choke, four-speed transmission, foot pedal throttle and brake.

SAFETY AND STABILITY —welded tubular steel frame, all steel body. Tandem seating maintains center of gravity on longitudinal axis, insures maximum stability under all conditions.

MESSERSCHMITT

MESSERSCHMITT

*Easy to own, economical to operate,
parts and service available at all dealers.
Check these big-car features . . .*

SPECIFICATIONS — STANDARD MODEL KR-175

Engine

Type	Fichtel & Sachs Single-cylinder air-cooled two-stroke engine
Bore	2 7/16 in.
Stroke	2 9/32 in.
Piston Displacement	10.6 cu. in.
Compression Ratio	6.6:1
Spark Advance	13/64 in. (5 mm.) B.T.D.C.
Carburetor	Bing Model 1/24 piston-valve carburetor with wet air cleaner and choke; slow-running and main jets accessible from outside.

Power Transmission

Type	Four-speed transmission, forward and reverse
Clutch	Three-disc cork-lined
Gear Ratios	Crankshaft to intermediate shaft, 2.12:1 — Intermediate shaft to main shaft: 1st gear, 3.22:1 — 2nd gear, 1.85:1 — 3rd gear, 1.24:1 — 4th gear, .95:1
Final Drive	1/2" x 5/16" roller chain driving rear wheel
Starter	Push button electric starter

Electrical Equipment

Ignition	Bosch 12-volt, 45-watt flywheel (generator)
Spark Plug	Bosch M-175-T-11 (for heavy-duty service, Bosch M-225-T-11)
Head Lights	4 1/8 in. (105 mm.) diameter, 25 + 25-watt double-filament light bulbs
Tail Lights	Double number plate illumination; tail lamp, cat's-eye reflectors, brake lamp
Direction Signal	Blinker lights with pilot lamp
Horn	Electric horn
Key	Removable light and ignition key
Battery	2 6-volt, 6.9 amp-hours (Batteries connected in series)

Body

General	All-steel body of streamlined design for minimum drag
Hood	Seamless deep-drawn Plexiglas hood
Windows	Large-size Plexiglas sliding windows on either side (removable); safety glass windshield
Luggage Compartment	Rear luggage compartment,
Seats	Convenient hinged front seat; back seat removable for additional luggage space; both seats upholstered with rubber-impregnated horse-hair, with foam rubber pads.

Chassis Frame

Frame Structure	Precision steel tubular construction of all-welded design
Steering Mechanism	Front wheels steered by steering knuckles; turning radius 13 ft.
Tires	4.00 — 8 (all wheels interchangable)
Tire Pressure (Air)	Front, 10 1/2 lbs./sq. in. rear, 21 1/2 lbs./sq. in.
Wheel Suspension	Front wheels independently suspended by three-directional rubber mountings
Brakes	Internal-shoe drum brakes, operated by steel cable; foot-pedal and self-locking hand brake acting on all three wheels

Dimensions and Weights

Overall Dimensions	Length 111 in.; width 48 in.; height 47 in.; wheelbase 80 in.; track 36 1/4 in.
Ground Clearance	6 1/4 in.
Weight	485 lbs.
Tank Capacity	3 gallons — (gasoline)/oil mixture 1:25; reserve tank capacity: 1 1/2 quarts
Engine Capacity	9 HP (continuous) at 5250 r.p.m., 9 1/2 HP (maximum) at 5250 r.p.m.
Fuel Consumption	100 mi./gallon
Speed	Maximum speed approx. 60 m.p.h.; continuous or cruising speed 50 m.p.h.; high performance on hills
Standard Equipment	Speedometer, windshield wiper, rear-vision mirror, complete set of tools, spare wheel and tire

SPECIFICATIONS CHANGEABLE WITHOUT NOTICE

Radio and other equipment available as optional accessories

For more information contact.

THE MESSERSCHMITT KABINENROLLER

Ingenious German Three-Wheeled Cabin Scooter Reveals Marked Performance Abilities

DURING the war, most of us were far too familiar with the products of the Messerschmitt factory. Now, Willy Messerschmitt has turned his swords into ploughshares, and has started to manufacture very small cars of extremely unconventional design. Their aircraft ancestry is apparent in every line, and the little machines have created immense interest. I decided, therefore, that a change from the super-speed models which I generally handle would not come amiss, and I travelled down to Worthing where Rudds, the distributors, had offered to lend me a "Kabinenroller".

That name means, literally, "cabin scooter", which explains the purpose of the Messerschmitt. It provides, in fact, the same economical transport for two that the best motor scooters give, but with the comfort and weather protection of a saloon car. Added to this is an astonishingly lively performance and much more speed than one would consider possible. Many owners of powerful sports cars have been looking for just such a vehicle as this, for nobody likes wasting a thoroughbred machine on trivial local journeys.

The basis of the Messerschmitt is a triangulated tubular frame, to which the steel body panels are welded, and form a stressed skin. Tandem seating has been chosen to make a slim silhouette possible, for streamlining is relied upon to ensure a high maximum speed on low power. The passenger's legs are on either side of the driver's seat, and there is ample room for two big people. The top of the body is of transparent plastic, except for the actual screen, which is of normal safety glass. The appearance is very reminiscent of the cockpit of a fighter aircraft.

A bulkhead behind the passenger's seat insulates the body from the engine compartment. This contains a Sachs two-stroke motor in unit with a four-speed gearbox, with air-cooling by forced draught. The blower rotor forms part of the combined dynamo and starter, which is a very neat assembly. The engine embodies reverse flow scavenging and a light alloy head. A chain drives the

PLOUGHSHARE: The Kabinenroller, comfortably accommodating Bolster and passenger, is a far cry from wartime Messerschmitt products, yet reveals some notable aircraft characteristics in design.

single rear wheel, and the suspension is by trailing forks and a bonded rubber springing unit.

Bonded rubber springs are also used in front. The i.f.s. is by short swing axles, and the two front wheels are steered through a divided track rod. There is no reduction gear in the steering linkage, and a pair of handlebars replaces the usual wheel. The left hand operates a twist grip throttle control, but the clutch and brakes have normal car-type pedals. There are cable operated brakes on all wheels.

The Kabinenroller is easy to enter, for the whole top of the body opens sideways. The driver's seat moves up and back on a linkage, to allow one to

get in without acrobatics. With the top closed the little machine becomes as cosy and weatherproof as any car. Large sliding windows ensure adequate ventilation.

The starter is extremely powerful, and the engine bursts into life at once whether it is hot or cold. The gears are engaged on the positive stop principle. One moves the right-hand lever forward to change up, and back to select a lower gear; the movement is, in fact, identical to that of a Cooper 500. The clutch is a little on the fierce side, but a smooth start can be made after some practice. A fairly deliberate movement of the lever is best, to avoid any risk of missing a gear. A small separate control, which selects

"The engine is easily the best two-stroke I have ever come across". (Above) The 174 c.c. Sachs motor is installed in conventional motor-cycle style.
(Left) "No normal driver need ever fear that he will tip up his cabin scooter". Only by excessively fast cornering did Bolster manage to lift an inside front wheel.

neutral from any gear, is a useful refinement.

The engine is easily the best two-stroke I have ever come across. It is entirely free of four-stroking or any other vices, and pulls strongly from quite low speeds. It will run smoothly right up to 7,500 r.p.m. I averaged 85 m.p.g. on short journeys, and I think one could approach 100 m.p.g. when cruising steadily on a long run. There is no pinking on lower-grade petrol. The speed of 62 m.p.h. is far ahead of any other miniature car's performance, and on only 174 c.c. it really is beyond all praise.

The ride is fairly hard, but quite reasonably comfortable. The wheels stick down to the road very well over all surfaces, and the brakes are extremely powerful. One can corner quite fast, and under these conditions all three wheels leave black lines on the road. I did succeed in lifting the inside front wheel, but only by cornering at an insane speed on a reverse camber. No normal driver need ever fear that he will tip up his cabin scooter.

A feature that merits some discussion is the direct steering. One presumes that it has been chosen to appeal to those who have graduated from scooters or autocycles. For a car driver, though, it is much too quick, and it took me about 10 miles before I could hold the machine really straight at maximum speed. I would suggest a geared steering, perhaps with a wheel instead of handlebars, to be offered as an extra for car folk. The present steering is entirely accurate and the driving position is very comfortable, but I am sure that many prospective owners would prefer something less direct.

On a journey, one can put up a surprisingly good average speed. Driving gently, a conversation can be carried on in normal tones, and the car is not noisy. If the willing little motor is given its head, though, a pretty powerful booming sound blots out all but the loudest speech. For those who wish to whisper sweet

ACCELERATION GRAPH

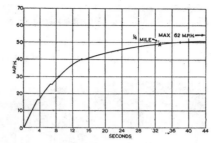

nothings, therefore, the cruising speed must be moderated accordingly.

The Messerschmitt is really beautifully made, and the appearance is most attractive. The interest aroused is immense, and if one pulls up outside a hostelry, the bar will empty in an instant. Everybody crowds round, and one soon finds oneself giving trial runs to all and sundry. Except for the warmth and comfort, it would easily be possible to forget that there was a roof at all, so light and airy is the interior. My passengers all remarked on this, and the cabin did not become objectionably hot, as one might expect. Among the extras available is a blind for very sunny weather, but this is not normally necessary.

Other optional equipment includes, believe it or not, a built-in radio. External luggage racks are also supplied if required. There is quite a useful parcel space behind the passenger compartment, and the rear seat can be instantly removed for the carriage of very bulky objects.

I enjoyed using the Messerschmitt enormously, and covered quite a large mileage with it. It is the kind of little machine that inspires an amused affection, and my children always referred to it as the Easter Egg. The look of amazement on the faces of many of the drivers that I passed on hills was always good for a laugh. Shouting and cheering often

broke out spontaneously from the pavements, and some people stood in an obvious daze, being apparently unable to believe what they had seen.

I was very impressed with the powerful lights, which gave a brilliant white beam. The flashing direction indicators were a useful and unexpected feature, too. There is, of course, no reverse gear, but this is seldom an inconvenience. I would call this an ideal shopping vehicle, and it is certainly a most practical means of transport.

The Messerschmitt Kabinenroller is a remarkable technical achievement. With its very advanced streamlining and super-efficient engine, it produces more economical speed than any other small car in the world. It is so well made that a long, trouble-free life is assured. Finally, the speedometer of the machine I tested was completely accurate, a virtue so rare as to be worthy of special remark.

SPECIFICATION AND PERFORMANCE DATA

Car Tested: Messerschmitt "Kabinenroller" 2-seater coupé. Price £274 15s., plus £61 3s. 2d. P.T.

Engine: Fichtel und Sachs single-cylinder air-cooled two-stroke, 62 mm. x 58 mm. (174 c.c.), 9 b.h.p. at 5,250 r.p.m., 6.6 to 1 compression ratio. Bing carburetter. Siba coil ignition.

Transmission: By short enclosed chain to cork-lined three-disc clutch. Four-speed gearbox, with right-hand control, ratios 4.8, 6.2, 9.3, and 16.2 to 1. Final drive by semi-enclosed chain to single rear wheel.

Chassis: Triangulated tubular frame with steel body panels forming stressed skin. Independent front suspension by swing axles and rear suspension by trailing radius arms, bonded rubber units in compression all round. Interchangeable bolt-on steel disc wheels, fitted 4.00 x 8 in. tyres. Cable operated brakes on all wheels with 4½ ins. x 1¼ ins. drums.

Equipment: 12-volt lighting and starting. Flashing direction indicators, speedometer. hand-operated windscreen wiper.

Dimensions: Wheelbase, 6 ft. 8 ins. Track, 3 ft. 0¼in. Overall length, 9 ft. 3 ins. Overall width, 4 ft. Height, 3 ft. 11 ins. Ground clearance, 6¼ ins. Turning circle, 26 ft. Weight, 385 lb.

Performance: Maximum speed 62 m.p.h. Speeds in gears, 3rd 40 m.p.h., 2nd 27 m.p.h. Standing quarter-mile 32.5 secs., 0-30 m.p.h., 8.2 secs.; 0-40 m.p.h., 14.4 secs.; 0-50 m.p.h., 38 secs.

Fuel Consumption: Driven hard, 85 m.p.g.

The 174 c.c. Two-stroke Sachs-powered Three-wheeler

MESSERSCHMITT

An Enclosed "Kabinenroller" Built by a Famous German Aircraft Factory and Having an Excellent Performance

(Above) Despite its narrow track, the Messerschmitt cornered well, and proved to be a suitable vehicle for town use. Note the excellent streamlining apparent in this view.

(Right) The engine room—showing the fan cooling of the 174 c.c. Sachs two-stroke; rear suspension pivot point concentric with the gearbox sprocket; and the partial chain enclosure.

TESTER'S ROAD REPORT

Maximum Speeds in:—

			Time from Standing Start
Top Gear (Ratio 4·8 to 1)	56 m.p.h.	= 5.800 r.p.m.	50 secs.
Third Gear (Ratio 6·2 to 1)	47 m.p.h.	= 6.200 r.p.m.	27·8 secs.
Second Gear (Ratio 9·3 to 1)	34 m.p.h.	= 6.700 r.p.m.	13·5 secs.

Speeds over measured Quarter Mile:—

Flying Start ...54·6... m.p.h.   Standing Start ...31·7... m.p.h.

Braking Figures On DRY TARMACADAM Surface, from 30 m.p.h.:—

ALL ~~Both~~ Brakes 32 ft. ~~Front Brake~~ ft. ~~Rear Brake~~ ft.

Fuel Consumption:—

30 m.p.h. 128 m.p.g.   40 m.p.h. 105 m.p.g.   50 m.p.h. 85 m.p.g.

Cabin accommodation for two—plainly depicting the main controls and clearly illustrating the manner in which the cabin top hinges for entry and egress.

WHEN analysing the design philosophy underlying the creation of the Messerschmitt "Kabinenroller," it is best to consider the vehicle not as a three-wheeled car —the conventional type of tricycle—but as a motor scooter (offering more than the average in the way of luxury) to which automatic stability has been imparted by the use of an extra wheel.

This, indeed, is the purpose of the design, and right well does it fulfil it, as the model recently tested by *Motor Cycling* demonstrated most convincingly. During the 12 days in which it was " on charge " at Bowling Green Lane, the " Kabinenroller " covered over 1,200 miles, some hundreds of which were taken "two-up." This distance included a 60-mile return trip every day of the working week, besides general about-town running and week-end touring. During this period, the only routine jobs carried out consisted of twice inflating the tyres, tightening one grub-screw and taking up the brake adjustment on three occasions. Nothing else required attention, and no trouble whatsoever was experienced.

Before going into details of the performance, however, a short technical description is necessary. Basis of the " Kabinenroller " is a duplex tubular frame, of welded-up construction, carrying at the forward end the twin front wheels, which are independently sprung on rocking arms, the movement of which is in each case controlled by a graduated rubber spring. Steering is by direct linkage from a steering column, equipped with handlebars. At the rear is attached a separate sub-frame—with a three-point rubber mounting—which carries the 174 c.c. Sachs fan-cooled two-stroke engine (fitted with a Siba dynamotor starter). Rear suspension, again rubber controlled, is by means of a swinging arm, which comprises, also, almost total enclosure of the final-drive chain. The pivot points of the arm are so arranged that constant chain tension is achieved.

Upon this " chassis " are built up the pressed-steel nosepiece, side panels, scuttle and rear " bonnet." As is to be expected in a design originated by Prof. Willy Messerschmitt, the streamlining is distinctive.

Tandem seating for the driver and passenger has allowed a comparatively narrow " fuselage," with three-dimensional streamlining. The nosepiece is well blended into the aircraft-type Plexiglass canopy (which incorporates a shatter-proof front screen of " Sekurit " glass), and the tail section has a pronounced downwards sweep. Equal care has been lavished on the important undersection, which is curved and carefully faired as far as the rear of the cabin section. Wide, smooth-contoured " wings " cover the front wheels; the headlamps are neatly faired into the nose, the twin tail lamps being equally unobtrusively blended into the rear.

Two car-type seats are provided in the " cockpit," that for the driver being coil-spring suspended. Controls take the form of a twist-grip throttle on the left handlebar; positive-stop gear control by lever on the right of the driving seat; ratchet-type hand brake; car-type clutch and brake pedals. The dashboard contains the ignition switch, with detachable haft; main beam indicator light; speedometer; " winker " switch; starter button and ignition warning light. The fuel tap is remotely operated by a lever on the rear bulkhead, and there is also a remote control for the float-chamber tickler. Accessibly positioned in the nose, to the driver's left, is a neat fusebox for the electrical system, much of the wiring for which is channelled through the inner chassis tubes.

Getting In

Access to the cockpit is gained by hinging to the right the complete scuttle/hood assembly, which is retained in the open position by a leather strap. A clamp-lock secures it when closed. On the early model tested, rear accommodation was somewhat cramped, but this point has been rectified on later production versions. In both cases, the driver has adequate leg-room.

Even on a cold morning, after the " Kabinenroller " had spent a night in the open, starting procedure was simplicity itself. With ignition and fuel " on," the tickler operated for a few seconds, and the starter button pressed. Immediately, the engine responded, ticking over slowly and evenly. First-gear selection was invariably positive. Equally invariably, clutch take-up was on the fierce side.

Initially, the experienced motorcyclist might have some difficulty in mastering the left-hand twist-grip throttle control. For this reason, it was found advisable to steer with the left hand only until one had become thoroughly accustomed to the new procedure.

Once this had been done, it was easy to make snappy gear changes, the positive-stop gear lever being ideally situated. If in a hurry, the drill was to " wind 'er up " in third gear, delaying the change until the speed was well into the forties. For more " pedestrian " motoring, it was quite a practical proposition to have top gear home by the time the needle had reached the 20-m.p.h. mark.

Even a eulogy would not be praise enough for the ever-willing Sachs engine, which never missed a single beat in the whole test. It seemed content to run on next to no fuel at all; it was satin-smooth in operation; it pulled well; it was easy to start. What more could one want . . . especially when it provided a top speed which was more than respectable for so small a vehicle?

Though the suspension seemed to be somewhat hard, the Messerschmitt handled well, if tending to become a little skittish on rippled surfaces at speeds of more than 50 m.p.h. Thanks to the direct steering, however, the driver was always in command of the machine; any tendency to " break away " on fast corners could be felt immediately, and appropriate action taken. Actually, despite its comparatively narrow track and long wheelbase, the " Kabinenroller " was surprisingly stable and could be hauled around corners in a manner which suggested that the makers had incorporated a pot of glue in each wheel! True, the tester *was* able to lift one of the front wheels—but only by deliberately attempting to do so by cornering hard at 30 m.p.h. with the machine

fully loaded. In general use, no private owner would be likely to do so.

As a serious means of transport the " Kabinenroller " commanded respect. A London—Hove trip of 56 miles was, on one occasion, packed in to just a fraction under 1 hr. 40 min.; a Brighton—Ashford journey of 60 miles (and hilly ones, too) occupied only a couple of hours; the subsequent Ashford—London trip took some 15 minutes less, despite rush-hour traffic jams.

It was here, of course, that the Messerschmitt—with its narrow track and good forward visibility—scored heavily. It could be directed instantly into any gap which presented itself, and its handiness when conditions were bad was almost equal to that of a solo motorcycle. On the version tested, the edges of the front wings were just out of view, which made things a little tricky, but on later models the front seat has been re-positioned, with consequent improvement.

As an economical runabout the " Kabinenroller " exceeded all expectations, an overall fuel-consumption figure of 90 m.p.g. being recorded, even when the model was cruised at more than 50 m.p.h. on open roads for long periods.

On hills, whether one-up or with a passenger, it had power in hand—when the throttle was tweaked upon smartly in bottom gear on a gradient of almost 1 in 5 the fully laden three-wheeler very nearly took off!

Weather protection was, of course, excellent and the cockpit was warm, even on the coldest mornings. As the sun rose, the temperature inside did so too, but it could be controlled by opening the sliding windows which—for summer conditions—are fully detachable.

Noise was one drawback of the test machine, but in a later model which was driven afterwards, all but a degree of " window rattle " had disappeared. Detail points which caused a few hard thoughts were the necessity for raising the rear bonnet to refuel. This, in turn, meant that the cabin had to remain open—not pleasant in heavy rain. A further point arising from this was that it necessitated the emptying of the parcel tray atop the bonnet before each refilling of the tank.

Much driving was done at night, and here the comprehensive " electrics " proved their worth—headlamps, sidelights, winking indicators with dashboard warning, tail lamps and stoplight all functioned well. With such an array, it was a little surprising to find that the windscreen wiper was manually operated from a small handlebar lever.

Unusual in layout, revolutionary, perhaps, in conception, the Messerschmitt " Kabinenroller " proved itself to be a thoroughly practical vehicle, possessed of a degree of economy and performance far above the average.

BRIEF SPECIFICATION

Engine: Sachs two-stroke, 62 mm. bore by 58 mm. stroke=174 c.c.; cast-iron cylinder barrel; light alloy cylinder head; C.R. 6.6 to 1; forced cooling from flywheel fan, with light alloy trunking.
Gearbox: Sachs four-speed in unit with engine; positive-stop, hand control; multiplate clutch; ratios 4.8, 6.2, 9.3 and 16.2 to 1; neutral-finding device.
Lighting and Ignition: Siba 12-volt, 75-watt flywheel dynamo and self-starter, with cooling fan; coil ignition; twin batteries;

twin headlamps; twin tail lamps; stop lamp; electric horn; winking indicators.
Chassis and Body: Welded-up, tubular steel chassis with steel panelling; car-type seating; plastic cabin with " Sekurit " glass windscreen; front and rear suspension of swinging type, rubber control'ed.
Wheels: Pressed-steel, with 4.00 x 8-in. Michelin tyres.
Tank: Petroil tank 2.4 gal. (approx.).
Dimensions: Wheelbase, 80 in.; track, 36¼ in.; height, 47 in.; ground clearance 6¼ in.;

overall length, 111 in.; weight, 460 lb. (approx.).
Price: £274 15s. plus £61 3s. 2d. Purchase tax=£335 18s. 2d.
Extras: Spare wheel, £5; chromed hub caps, 18s. 3d. each; clock, £3 1s. 7d. plus 10s. 4d. P.T.; luggage rack, £2 2s. (£3 3s. chromed).
Manufacturers: Regensburger Stahl und Metallbau G.m.b.H., Regensburg, Germany.
Concessionaires: Beulah Hill Eng. Co., Ltd., 411 Beulah Hill, London. S.E.19.

MESSERSCHMITT CABIN SCOOTER
Specification of Model KR-175

ENGINE
Manufacturer	Fichtel & Sachs
Type	Single-cylinder air-cooled two-stroke engine
Piston displacement	10.6 cu.in. (174 c.c.)
Compression ratio	6.6:1
Carburettor	Bing model 1/24 piston-valve carburettor with wet air cleaner

POWER TRANSMISSION
Type	Four-speed transmission, hand-operated
Clutch	Three-disc cork-lined
Starter	Electric

CHASSIS FRAME
Frame structure	Precision steel tubular construction of all-welded design
Steering mechanism	Front wheels steered by steering knuckles; turning radius 13 ft.
Tyres	4.00-8 (all wheels interchangeable)
Wheel suspension	Front wheels independently suspended by three-directional rubber mountings.

ELECTRICAL EQUIPMENT
Ignition & starter Bosch 12 volt; 75 watt flywheel dynamo & starter, headlamp, tail lamp, direction indicators, horn and battery.

BODY
General	All-steel body of streamlined design for minimum drag
Luggage compartment	Rear luggage compartment, capable of being locked.

DIMENSIONS & WEIGHTS
Overall dimensions	Lgth.111 in; width 48in; height 47 in; wheelbase 80 in; ground clearance 6¼ in.
Weight	335 lbs
Tank capacity	2½ Imp.gall. petrol/oil mixture; 1:25; reserve tank 1.1/3 Imp.qt.
Engine capacity	9HP (continuous) at 5250 rpm; 9½ HP (max) at 5250rpm
Fuel consumption	123 miles per gallon
Speed	Max.speed approx. 60 mph continuous or cruising speed 47 mph
Standard equipment	Speedometer, windscreen wiper, rear-vision mirror, complete set of tools
Special equipment	Spare wheel, chrome-plated hub caps, clock and luggage rack.

PRICE RETAIL (ex.works) ... £248. 7. 0. PURCHASE TAX ... £51. 8. 0.

Specification and price subject to change without notice

£74.15. 0. DOWN ---- £11. 5. 0. per month over 24 months.

MESSERSCHMITT
"CABIN SCOOTER"

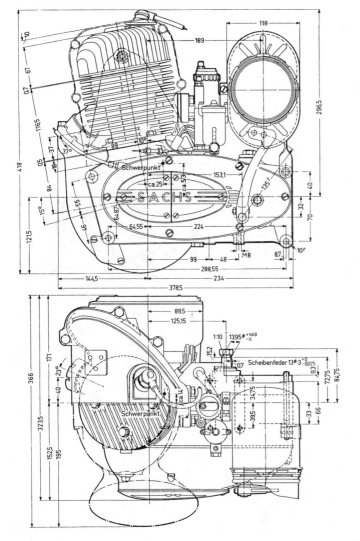

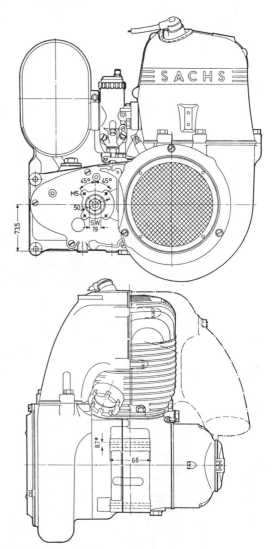

Im Interesse der konstruktiven Weiterentwicklung bleiben Änderungen vorbehalten

The MESSERSCHMITT IS "TOPS"

Two concentrated, piercing shafts of light, the two headlamps of the Messerschmitt Three-Wheeler penetrate even the darkest night. The two passengers sitting snug beneath the elegant allround

visibility hood do not mind the cold airstream whistling past outside. Neither cold nor rain can worry them. The slim, streamline body of their vehicle cuts safely and speedily through the narrow, dark side streets at night and just as easily threads its way through the densest city traffic. Whereas the "big fellows" have to search desperately for a parking space large enough to accommodate their bulk, a three-wheeler can squeeze in just anywhere. And the Reception Clerk at the desk of the Grand Hotel is happy to welcome such cheerful and well-dressed guests whose choice of car proves their progressive outlook.

No matter whether you are driving to work or setting out for your holiday: The MESSERSCHMITT IS "TOPS".

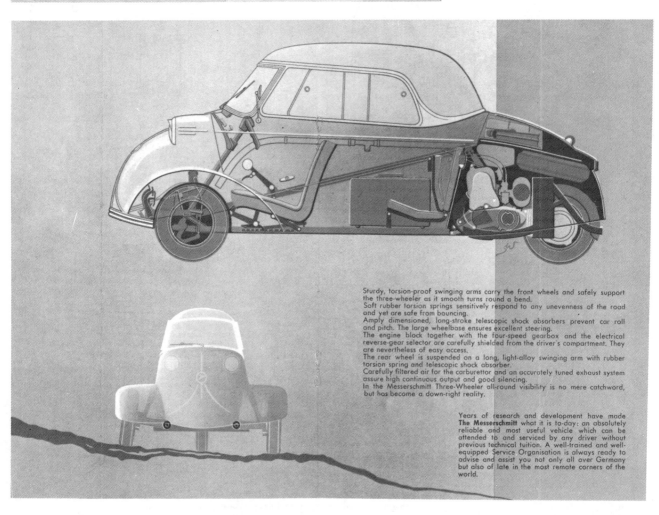

Sturdy, torsion-proof swinging arms carry the front wheels and safely support the three-wheeler as it smooth turns round a bend.
Soft rubber torsion springs sensitively respond to any unevenness of the road and yet are safe from bouncing.
Amply dimensioned, long-stroke telescopic shock absorbers prevent car roll and pitch. The large wheelbase ensures excellent steering.
The engine block together with the four-speed gearbox and the electrical reverse-gear selector are carefully shielded from the driver's compartment. They are nevertheless of easy access.
The rear wheel is suspended on a long, light-alloy swinging arm with rubber torsion spring and telescopic shock absorber.
Carefully filtered air for the carburettor and an accurately tuned exhaust system assure high continuous output and good silencing.
In the Messerschmitt Three-Wheeler all-round visibility is no mere catchword, but has become a down-right reality.

Years of research and development have made The Messerschmitt what it is to-day: an absolutely reliable and most useful vehicle which can be attended to and serviced by any driver without previous technical tuition. A well-trained and well-equipped Service Organisation is always ready to advise and assist you not only all over Germany but also of late in the most remote corners of the world.

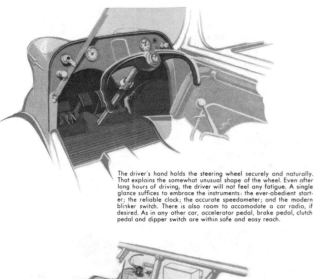

The driver's hand holds the steering wheel securely and naturally. That explains the somewhat unusual shape of the wheel. Even after long hours of driving, the driver will not feel any fatigue. A single glance suffices to embrace the instruments: the ever-obedient starter; the reliable clock; the accurate speedometer; and the modern blinker switch. There is also room to accomodate a car radio, if desired. As in any other car, accelerator pedal, brake pedal, clutch pedal and dipper switch are within safe and easy reach.

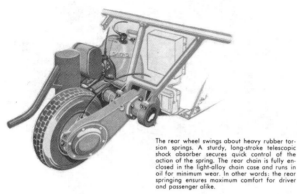

The rear wheel swings about heavy rubber torsion springs. A sturdy, long-stroke telescopic shock absorber secures quick control of the action of the spring. The rear chain is fully enclosed in the light-alloy chain case and runs in oil for minimum wear. In other words: the rear springing ensures maximum comfort for driver and passenger alike.

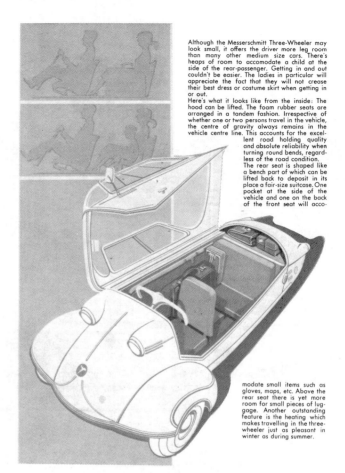

Although the Messerschmitt Three-Wheeler may look small, it offers the driver more leg room than many other medium size cars. There's heaps of room to accomodate a child at the side of the rear-passenger. Getting in and out couldn't be easier. The ladies in particular will appreciate the fact that they will not crease their best dress or costume skirt when getting in or out.

Here's what it looks like from the inside: The hood can be lifted. The foam rubber seats are arranged in a tandem fashion. Irrespective of whether one or two persons travel in the vehicle, the centre of gravity always remains in the vehicle centre line. This accounts for the excellent road holding quality and absolute reliability when turning round bends, regardless of the road condition.

The rear seat is shaped like a bench part of which can be lifted back to deposit in its place a fair-size suitcase. One pocket at the side of the vehicle and one on the back of the front seat will acco-

modate small items such as gloves, maps, etc. Above the rear seat there is yet more room for small pieces of luggage. Another outstanding feature is the heating which makes travelling in the three-wheeler just as pleasant in winter as during summer.

Technical Data KR 200

Engine 200 cc. fan-cooled single-cylinder F.&S. two stroke engine. Powerful acceleration, high performance on hills, higher output. Especially high engine torque with l o w r.p.m. Reverse gear to be electrically selected. Rubber mounted rear wheel suspension, which is separated from the engine. Intermediate power transmission by cardan shaft. Cylinder bore: 2.5 ins. (65 mm), Piston stroke: 1.98 ins. (58 mm), Cylinder capacity: 191 c.c., Compression ratio: 6,6:1, Carburettor: Bing piston valve carburettor. Electric starter: 12 volt.

Gears, Transmission

Type: Four-speed gearbox, hand-operated; accelerator, clutch, brake and traffic beam switch are operated by footpedals. Noiseless gear shifting by Teleflex cable.

Final drive: Roller chain ¹/₂ × ¹/₄" driving rear wheel, chain in oil bath.

Chassis

Frame: Torsion-proof steel tube frame with completely closed floor.

Steering mechanism: Front wheels steered by divided track rods, rubber mounted. Modern steering bar with horn button.

Suspension: Each wheel separately suspended with enlarged track width, soft rubber torsion springs with hydraulic shock absorbers on all three wheels.

Body: All-steel body with very durable furnace-set coat of paint, Large sliding side windows; Panoramic windscreen of security glass. Full-view dome, convertible perspex top (cabrio-limousine). Electric wind screen wiper. Baggage compartment in rear hood.

Tyres: 4.40 × 8 with special surface, all wheels interchangeable.

Brakes: Internal expanding brakes activated by cables, foot brake and self-locking handbrake acting on all 3 wheels.

Cockpit: Seats (one behind another) with foam rubber upholstery. Front seat adjustable which can be tipped up for convenient entrance, full width back seat to fold up. Accomodation for an adult and a child.

Measurements and Weights: Length 111 ins.
Width 48 ins.
Height 47 ins.
Wheelbase 80 ins.
Track width 42¹/₂ ins.
Ground clearance 6¹/₂ ins.
Weight 463 lbs.
Pay load 403 lbs.

Tank capacity: 3 gallons petrol/oil mixture, of which ¹/₂ gallon is reserve.

Performance: 10 h.p. (B.H.P.) at 5250 r.p.m.

Fuel consumption: 100 m.p.g.

Speed: Maximum 62 m.p.h., cruising 53 m.p.h.

Electrical Equipment

Ignition System: Coil.

Lighting System: 90 watt, 12 volt.

Tail Lights: Two-side illumination for number plates and rear light, reflector, stop light, flashing type direction indicator on both sides and in the rear..

Batteries: 2 × 12 Ah/6 V, 12 Amp. series-connected.

Specifications to be changed without notice.

THE GORDON MOTORS CORP. Yonkers, N. Y. Importer and Distributor For the U. S. A.

PRINTED IN WESTERN GERMANY

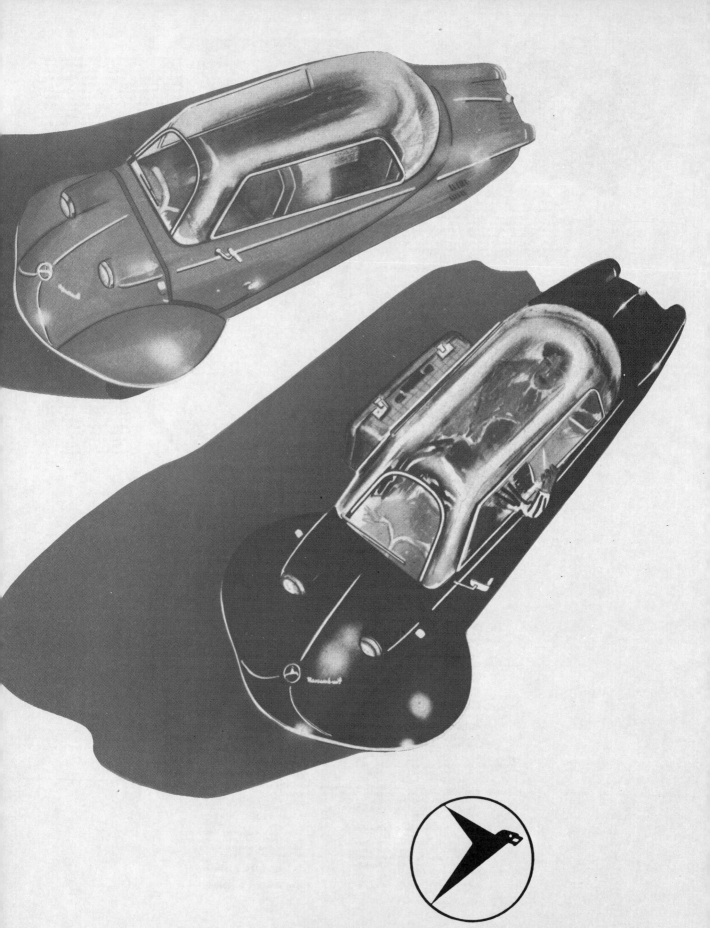

MESSERSCHMITT

MESSERSCHMITT KR 175 *Scootmobile*

De Driewielige Scooter met de unieke voordelen!

De grote voordelen:

Een volledige bescherming tegen alle weersinvloeden gaat gepaard met een vrij uitzicht naar alle zijden voor beide inzittenden door de van plexiglas vervaardigde kap. Twee comfortabele zittingen met rugleuningen. De zitplaatsen zijn doelbewust achter elkaar geplaatst, zodat ook solorijden de stabiliteit niet nadelig beïnvloedt. Electrische starter en volledige installatie met 12 V batterij. Voetkoppeling en vier versnellingsbak met eenvoudige neutraalstelling bij iedere versnelling. Zeer soepele voor- en achtervering. Bijzonder vaste ligging en stabiliteit in de bochten door laag liggend zwaartepunt en de plaatsing van twee wielen voor en een achter (hetgeen beter is dan het omgekeerde). Verbluffend acceleratievermogen en hoge gemiddelde snelheid. Verbazingwekkend goed klimvermogen. Beproefd op de hoogste Europese passen. Gestroomlijnde vorm, waardoor een zeer geringe luchtweerstand. Door de geringe breedte handig in het drukste verkeer, gemakkelijk te parkeren.

Technische gegevens:

Motor:

Constructie	2-takt, 1 cyl. geforceerde luchtkoeling
Motorvermogen	9 pk bij 5200 omw/min.
Boring	62 mm
Slag	58 mm
Cyl. inhoud	174 cm³
Compressie-verh.	6,6 : 1
Carburateur:	met gasschuif 1/24, stat. sproeier en olie-luchtfilter

Versnellingsbak:

Koppeling	met handbediening drie schijven met kurkbekleding
Overbr. verh.	van krukas naar hulpas 2,12 : 1 van hulpas tot hoofdas:
	1e versnelling 3,22
	2e „ 1,85
	3e „ 1,24
	4e „ 0,95

Overbrenging:

Rollenketting	1/2" x 5/16" op achterwiel

Starter Dynastart

Chassis:

Frame	gelaste stalen buizen van de beste kwaliteit, bodem uit één stuk geheel gelast
Stuurinrichting	op voorwielen door middel van fusee's, draaicirkel 7 m.
Banden	4.00 x 8, speciaal profiel, alle drie de banden en wielen zijn uitwisselbaar.
Vering	ieder wiel afzonderlijk wordt met een rubber bus geveerd.

Remmen	Trommelremmen door kabels bediend, hand- en voetrem werken op alle drie de wielen.

Electr. installatie:

Ontsteking:	12 V gelijkstroom installatie, bobine
Koplampen	diameter 105 mm, 25/25 W lampen
Starter	12 V Dynastart
Achterlicht	aan beide zijden met verlichting van de nummerplaat gecombineerd, achterlichten werken als reflectors, stoplicht in het midden.
Richtingaanw.	Knipperlichten met contrôle-lamp
Hoorn	Electrisch
Batterij	2 batterijen in serie, ieder 11 Ah.

Lakwerk:

	diverse kleuren leverbaar; uitgevoerd in prima lak

Carrosserie:

Kap	uit één stuk naadloos plexiglas geperst en gemakkelijk openklapbaar voor het instappen
Carrosserie materiaal	uit plaat vervaardigd, geheel gelast
Zijruiten	grote zijruiten, vervaardigd uit plexiglas de ruiten zijn schuifbaar en uitneembaar
Voorruit	van veiligheidsglas
Bagageruimte	afsluitbaar, boven motor, afmetingen ca. 50 cm diep, 60 cm breed en 17 cm hoog.

Zittingen	zeer comfortabele voorzitting, welke naar boven bewogen kan worden. Zitting voor passagier uitneembaar, waardoor eventueel extra ruimte voor bagage. Veel beenruimte.
Bekleding	van de zittingen bestaat uit schuimrubber, bekleed met kunstleer in bij het exterieur passende en aantrekkelijke kleur.

Afmetingen en prestaties:

Lengte	282 cm
Breedte	122 cm
Hoogte	120 cm
Spoorbreedte	92 cm
Wielbasis	203 cm
Vrije hoogte	16 cm
Draaicirkel	700 cm
Inhoud Tank onder motorkap	ca. 12 liter, benzine-oliemengsel. Verhouding 1 : 25 (1½ liter reserve)
Benzine-verbruik	ca. 40 km op 1 liter (2,3 tot 2,5 liter per 100 km) bij normaal gebruik.
Snelheid	max. snelheid 90 km/uur, kruissnelheid 75 km/uur (met ingelopen motor)
Wegenbelast.	f 42,— per jaar
Rijbewijs	„A" (voor motorrijwielen).
Standaard uitrusting:	snelheidsmeter, km teller, ruitenwisser, achteruitkijkspiegel, gereedschap, reservewiel met -band. Knipperlichten met contrôle lamp.

De fabriek behoudt zich te allen tijde het recht voor om veranderingen aan te brengen zonder voorafgaand bericht.

DE MESSERSCHMITT „SCOOTMOBILE"

BIEDT U:

— de bescherming en het comfort van een automobiel.

— de zuinigheid en eenvoud van een motorrijwiel.

● Reeds duizenden in Europa in gebruik.

    ● Het product van een fabriek met wereldnaam.

        ● Voor ons grillige klimaat een uitkomst.

            ● Geen speciale motorkleding nodig.

Importrice voor Nederland:

N.V. INTERNATIONALE AUTOMOBIEL MIJ
DEN HAAG - SCHELDESTRAAT 2 - TEL. 77.20.00

Voor detailverkoop:

Dealers door het hele land !

Roadster KR 201
MESSERSCHMITT

SPORTS CONVERTIBLE
DE LUXE

Reliable!

Exciting acceleration!

High cruising speed!

Most economical!

The vehicle for work and pleasure!

PLUS EXCEPTIONAL COMFORT!

FMR

FAHRZEUG- und MASCHINENBAU G.m.b.H.

REGENSBURG

GERMANY

Sports Convertible KR 201

The "MESSERSCHMITT" Sports Convertible has achieved such a perfection throughout the years that anybody, without any technical knowledge, can drive and service the "MESSERSCHMITT".

Specification may be changed without notice!

TECHNICAL DATA KR 201

Engine: SACHS 191 cc. Two Stroke; rubber mounted; Bore 65 mm (2.5 ins) Stroke 58 mm (1.98 ins); Compression ratio 6.6:1, Performance 9.7 BHP 5000 rpm; Bing piston valve Carburettor with special air intake silencer and air filter; special type Exhaust silencer for exceptional quiet running; Electric starter; Fan cooling; quick acceleration; efficient hill climbing; good turning circle.

Transmission: Four speed gear box; positive stop ratchet type; silent hand operated gear change by Teleflex control system; Electrical operated Reverse gear; Four plate clutch.

Final drive: From engine by Cardanshaft to rubber mounted pinion in independantly hung Rocker box.; Roller chain fully enclosed in oil bath and driving rear wheel.

Controls: Foot pedal operated Accelerator, Clutch, Foot Brake and Dipper Switch; Hand brake with ratchet and hand grip.

Frame: Torsion proof steel tube frame with completely closed floor.

Steering: Direct steering by wear-proof divided track rods; Modern steering bar with horn button.

Suspension: Each wheel independantly suspended, with enlarged track width; soft rubber torsion springs with hydraulic shock absorbers on all three wheels.

Tyres: 4.00×8. with all weather steel rims and ornamental wheel discs. Spare wheel (as extra) may be placed beneath the engine cover.

Brakes: Cable activated internal expanding foot brake operating on all three wheels; Selflocking hand brake.

Body: All steel body with very durable furnace-set coat of paint; Large removable pliable side curtains; Panoramic Windscreen of safety glass. Folding Canvas hood (Drop-head); Electric windscreen wiper; Lockable luggage compartment in rear; Safety lock for cabin; Sun Vizor and Luggage Carrier as extras.

Cockpit: Seats, one behind another (tandem) with foam rubber upholstry; Adjustable front seat can be tipped up for convenient entrance; full width back seat can be folded up; accomodation for two adults and a child.

Dimensions: Length 112 ins, Width 48 ins, Height 47 ins, Wheelbase 80 ins, Track width 43 ins, Ground clearance 6 ins, Unloaden weight 506 lbs, Pay load 440 lbs.

Performance: Maximum speed 70 mph; Climbing ability up to 1:3.

Electrical Equipment: 90-watt, 12 volt Lighting system; battery ignition; 2×35 watt headlamps; two-side illumination of number plates; twin rear lights with reflectors; Brake light; Blinkers operating in front and in rear lights; Blinker and beam indicator lights on dashboard. Ignition and reverse warning lights on dashboard.

Hailed by its makers as a utilitarian car, this vest-pocket job has wheelbase of 80 inches, 111-inch overall length.

Simplicity of operation and economy of use are stressed in Cruisette. Notice tilt top. Car uses cantilever steering.

Three Wheeler

The Messerschmitt

Cruisette is so

light you can ship

it as baggage!

AMERICAN motorists are getting a look at something new in the small-car field—the three-wheeled Messerschmitt Cruisette, so light in weight (484 pounds) that it can be shipped as baggage!

A product of Willy Messerschmitt, famed German aeronautical designer and industrialist, the two-seater Cruisette sells for $895 f.o.b. port of entry, federal tax included. Manufactured in Regensburg, Germany, it is hailed by Craighead-Koontz, Inc., of South Bend, Ind., American importers and distributors, as a utilitarian "second" car for suburban residents, or for use by college students and commuters.

Norman D. Craighead, of the South Bend firm, obtained one of the cars in Regensburg, drove it through the Bavarian mountains into Austria, through the Tyrol, into France and thence to Luxembourg, Belgium and Holland, where he put it aboard ship as baggage.

Arriving in New York, he drove it to South Bend, and from there to San Francisco and Los Angeles and back to South Bend, negotiating two mountain passes in weather that made page one of West Coast newspapers. The command to "fill 'er up" doesn't seriously crimp the pocketbook; 2.55 gallons are carried in the main tank, with .45 gallon in the reserve tank—and the manufacturers claim you can get from 65 to 80 miles per gallon!

Overall length is 111 inches, with an 80-inch wheelbase. Track is 36¼ inches, while the overall width is 48 inches, or an inch more than the height of the car.

The single-cylinder, two-cycle Fichtel & Sachs motor is forced air cooled. Of 10.6 cubic inch displacement, it has a compression ratio of 6.6:1.

Four speeds are provided in both forward and reverse. Internal expanding brakes are used on all three steel disc mounted wheels. The single rear wheel is used for power, with steering done by the two front wheels. Four-ply tires are 4.00x8.

A top speed of 55 to 60 mph. is claimed for the Cruisette, with a cruising speed of 45 to 50. The turning radius is 13 feet, 8 inches.   ☆

Tricycle test

the kabinroller

IN THE course of compiling road tests, usually a rather strained affair that results in more work than pleasure, we occasionally run across a machine that is really fun. The sort that you enjoy "playing with" and usually find yourself taking time from other duties to stretch the length of time on the road with the car. That was the case with the Messerschmitt, and we thoroughly enjoyed our session with the Kabinroller.

To begin with, the little car is startling in its appearance. It gives more the impression of an uncompleted airplane or a large insect, than a car or motor scooter. The plexiglas top swings up, to allow driver and passenger to clamber aboard and, when open, the tricycle looks very much like the Taylor Cub of a few years ago. Getting in, and folding our rather lengthy legs under the handlebars, while not graceful, was surprisingly easy due to the hinged, elevating front seat. The back seat is a bit more tricky, and although we had no opportunity to experiment with the theory, we feel that slacks would be almost mandatory

The "tricycle" features a deluxe interior and a two-tone color job.

Seat of the Messerschmitt raises to allow easy entry.

Handlebar steering gives the driver surprising control.

feminine attire for passengering in the car. Once aboard though, the seats, of foam rubber and hair, are most comfortable and the controls feel quite natural from the very first contact.

Starting the husky little Sachs engine is surprising. Even after the Messerschmitt has been sitting quite a while the power plant comes to life so quickly that rarely can you hear the starting motor spin it even one turn. When it catches too, there is a roar in the tight little cabin that would warm the cockles of the hearts of even devoted Ferrari addicts. This impression of power is pretty well founded too. The nine and one half horses push the car along quite respectably. In the de luxe model, the engine has an extra two horsepower to beef it up a bit, and will force a truly satisfying squeak from the rear wheel. Incidentally, the huskier version also has a rather novel system for reverse. The ignition key is turned to the reverse position (indicated by a colored light on the panel) and the whole engine restarts, running in the opposite direction.

Driving the little beast is very similar to driving a formula III car. The motorcycle type gear shift allows you to go up and down on the speeds with truly satisfying Grand Prix noises and style and coupled with the healthy sound effects from the engine gives the illusion of blistering performance. On the performance side, the acceleration is more apparent than real, giving a good 0-30 figure, about six seconds and very little after that. The feeling of urge that the little gadget has is very real and driving it in traffic, if you discount the incredulous stares from other motorists, is great fun.

As to other details of its operation, the Kabinroller gets a little bouncy around forty, feeling somewhat like a light plane taxiing across a dirt landing field, but it has very good directional stability and the bouncing isn't particularly bothersome. Top gear is a bit light for most driving except along the highway, and the car requires a bit of rowing to get it up hills when encumbered with two passengers but it will attain 60 fairly readily on the open road.

The only feature that we found disconcerting was the steering. The handlebar system, feeling for all the world like a motorcycle, is a bit too quick, and feels nothing at all like a wheel. At high speeds (above 45) the car has the feeling that it *could* get out of control without too much effort, but with practice the machine seems perfectly tractable.

One other point that deserves some mention is the

Plexiglas top lifts sideways for entry and clicks solidly into place.

heating-demisting system. There is a tube that pipes warm air into the cockpit on the same principle that the Volkswagen uses. This is loose at the front, normally fitting over a metal tube and directing the heat toward the driver's feet. To use it as a demister, the end of the tube is unplugged and hooked into a small padded hook on the instrument panel, thereby directing warm air onto the windscreen. It works beautifully.

As to the construction, it follows the accepted standard of German workmanship. The car seems really solid, and should be able to go the 100,000 miles with no exceptional difficulty.

In general the Messerschmitt is one of those happy little machines that combines ingenuity, economy and transportation with considerable amusement value and reliability. For under a thousand dollars you can't do better, and if you like gymkhanas, well . . .

* * *

Deluxe model engine is more powerful than standard version.

Spare tire is mounted on rear fender behind the gas tank.

SPECIFICATIONS—STANDARD MODEL KR-175

ENGINE

Type	Fichtel & Sachs Single-cylinder air-cooled two-stroke engine.
Bore	2 7/16 in.
Stroke	2 9/32 in.
Piston Displacement	10.6 cu. in.
Compression Ratio	6.6:1.
Spark Advance	

TRANSMISSION

Type	Four-speed transmission, forward and reverse.
Clutch	Three-disc cork-lined.
Ignition	Bosch 12-volt, 45-watt flywheel (generator).

DIMENSIONS AND WEIGHTS

Overall Dimensions	Length 111 in.; width 48 in., height 47 in.; wheelbase 80 in.; track 36 1/4 in.
Ground Clearance	6 1/4 in.
Weight	485 lbs.
Tank Capacity	3 Gallons—(gasoline)/oil mixture 1:25; reserve tank capacity 1½ quarts.
Engine Capacity	9 HP (continuous) at 5250 r.p.m.; 9½ HP (maximum) at 5250 r.p.m.
Fuel Consumption	100 mi./gallon.
Speed	Maximum speed approx. 60 m.p.h.; continuous or cruising speed 50 m.p.h.; high performance on hills.
Standard Equipment	Speedometer, windshield wiper, rear-vision mirror, complete set of tools, spare wheel and tire.

MESSERSCHMITT
KR 200 DE LUXE CABIN SCOOTERS

FAST
65 MPH
CRUISING SPEED
45-53 MPH

RELIABLE
STURDY ENGINE
ALLSTEEL BODY
TUBULAR STEEL FRAME

ECONOMICAL
90-100 MPG
RUNNING COSTS
LEES THAN 1d PER MILE

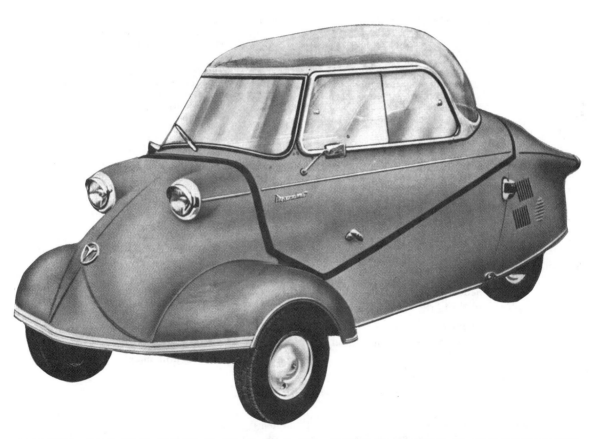

THE INCOMPARABLE MESSERSCHMITT
HIGH ACCELERATION - HIGHLY MANOEUVERABLE - EASY TO PARK
PERFECT ALLROUND VISIBILITY - PERFECT WEATHER PROTECTION
STREAM LINED BODY - COMFORTABLE SEATING FOR TWO ADULTS AND ONE CHILD

SOLE U.K. CONCESSIONAIRES: CABIN SCOOTERS (ASSEMBLIES) LDT. 80 GEORGE ST., BAKER ST.
LONDON W.I. HUNTER 0609

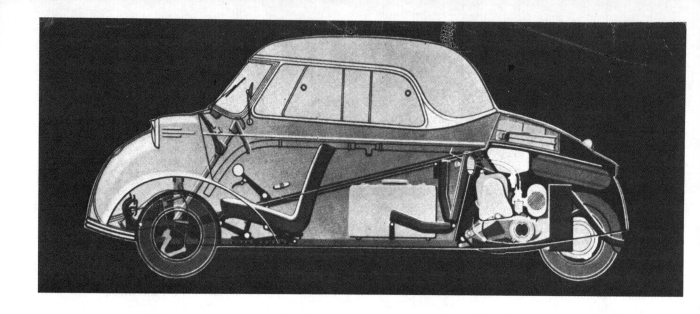

Technical Data KR 200

Engine 200 cc. fan cooled single - cylinder F. & S. two stroke engine. Powerful acceleration, high performance on hills, high output. Especially high engine torque with low r. p. m. Reverse gear electrically selected. Rubber mounted rear wheel suspension, which is separate from the engine. Intermediate power transmission by cardan shaft. Cylinder bore: 2.5 ins. (65 mm), Piston stroke: 1.98 ins. (58 mm), Cylinder capacity: 191 c. c., Compression ratio: 6,6:1, Carburettor: Bing piston valve carburettor.
Electric starter: 12 volt.

Gears, Transmission

Type: Four-speed gearbox, hand-operated; accelerator, clutch, brake and traffic beam switch are operated by footpedals. Noiseless gear shifting by Teleflex cable.

Chassis

Frame: Torsion-proof steel tube frame with completely closed floor.

Steering
mechanism: Front wheels steered by divided track rods, rubber mounted. Modern steering bar with horn button.

Suspension: Each wheel separately suspended with enlarged track width, soft rubber torsion springs with hydraulic shock absorbers on all three wheels.

Body: All-steel body with very durable furnace-set coat of paint, Large sliding side windows; Panoramic windscreen of safety glass. Full-view dome or convertible top (cabrio-limousine). Electric wind screen wiper. Baggage compartment in rear hood.

Brakes: Internal expanding brakes activated by cables, foot brake and self-locking handbrake acting on all 3 wheels.

Cockpit: Seats (one behind another with foam rubber upholstery. Front seat adjustable which can be tipped up for convenient entrance, full width back seat to fold up.

Measurements and
Weights: Length 111 ins.
Width 48 ins.
Height 47 ins.
Wheelbase 80 ins.
Track width 42 $^1/_2$ ins.
Ground clearance 6 $^1/_2$ ins.
Weight 463 lbs.
Pay load 403 lbs

Tank
capacity: 3 gallons petrol oil mixture, of which $^1/_2$ gallon is reserve.

Performance: 10 h. p. (B. H. P.) at 5250 r. p. m.

Electrical Equipment

Ignition
System: Coil.

Lighting
System: 90 watt, 12 volt.

Tail Lights: Two side illumination for number plates and rear light, reflector, stop light, flashing type direction indicator on both sides and in the rear.

Batteries: 2x12 Ah 6 V, 12 Amp. series-connected

Specifications to be changed without notice.

FMR

FAHRZEUG- UND MASCHINENBAU GMBH. REGENSBURG

YOUR LOCTE AGENT:

Der Messerschmitt-Kabinenroller ist in jahrelanger Ent-
wicklung zu einem unbedingt **zuverlässigen** Zweckfahr-
zeug herangereift, das von jedem Fahrer ohne besondere
technische Vorkenntnisse gepflegt und gewartet werden
kann. Eine gutgeschulte und vorbildlich eingerichtete
Kundendienstorganisation steht in jedem größeren Ort
mit Rat und Tat bereit.
Ihr Messerschmitt-Vertreter:

RSM MESSERSCHMITT-WERK · REGENSBURG

*Für jedes Wetter -
für jede Straße*

MESSERSCHMITT-KABINENROLLER

*D*er Messerschmitt-Kabinenroller wurde zum Symbol fort-
schrittlicher Fahrzeugtechnik:
Er vereinigt die Sparsamkeit des Motorrades mit dem
Temperament und der Wendigkeit eines Sportwagens,
die Anspruchslosigkeit und die ganz auf hohe Fahrleistung abgestimmte Zweckform der Fahrmaschine mit
der eleganten Linienführung und dem Fahrkomfort eines
Luxusgefährtes.

Schirm und Wetterkleidung? — Nein, Kabine!

Vom Flugzeug haben wir's gelernt:
Ein anatomisch richtig durchgebildeter, ermüdungsfreier
Bereitschaftssitz und volle Sicht
nach allen Seiten sind die
Grundlagen für die sichere
Beherrschung jeder
Verkehrssituation.

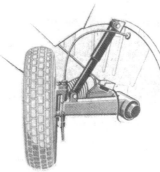

Kräftige Schwingen mit
wartungsfreier Gummi-
drehfederung und lang-
hubigen Teleskopstoß-
dämpfern bürgen für
beste Bodenhaftung,
hohe Spursicherheit und
angenehme Federung
bei allen Geschwindig-
keiten und auf allen
Straßen.

In zwei hintereinander
angeordneten Sitzen
liegt nicht nur das
Geheimnis für die un-
erreichte Windschlüp-
figkeit, sondern auch
die Zauberei mit dem
Raum: Der Fahrer des
Kabinenrollers hat eine
größere Ellenbogenfrei-
heit als sie ihm selbst
ein Wagen der mittle-
ren Klasse bieten kann.

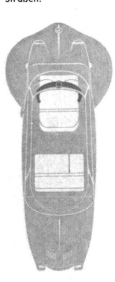

Messerschmitt-Kabinenroller KR 200

Motor	Gebläsegekühlter F. & S.-Einzylinder-Zweitaktmotor. Rascher Anzug, verblüffende Bergfreudigkeit, erhöhte Dauerleistung. Besonders hohes Drehmoment im niedrigen Drehzahlbereich. Gummigelagerte, vom Motor getrennt aufgehängte Hinterradschwinge mit Kraftübertragung durch Kardanzwischenwelle. Bohrung: 65 mm; Hub: 58 mm; Zylinderinhalt: 191 ccm; Verdichtung: 1:6,6; Leistung: 10 PS bei 5250 U/min. Bing-Startvergaser mit Naßluftfilter und Ansauggeräuschdämpfer.
Elektr. Anlasser:	12 Volt
Getriebe	Vierganggetriebe mit Ratschenhandschaltung. Geräuschlose Schaltung durch Teleflexzug. Elektrisch geschalteter Rückwärtsgang.
Antrieb:	Im Ölbad laufende ⅓ ¼″ Rollenkette auf Hinterrad.
Bedienungsorgane:	Fußgas, Fußbremse, Fußkupplung, Fußabblendschalter.
Fahrgestell	
Rahmen:	Verwindungsfreier Stahlrohrrahmen mit vollkommen geschlossener Bodenwanne.
Lenkung:	Direkte, verschleißfeste Achsschenkellenkung mit gummigelagerten Spurstangen. Moderner Lenker mit Signalknopf.
Federung:	Einzelradfederung mit verbreiterter Spur, weiche Gummidrehfedern mit hydraulischen Stoßdämpfern und sehr großen Federwegen an allen 3 Rädern.
Bereifung:	4,40 × 8″ Spezialprofil, sämtliche Räder untereinander auswechselbar.
Bremsen:	Durch Seilzug betätigte Innenbackenbremsen. Fußbremse und feststellbare Handbremse auf alle 3 Räder wirkend.
Karosserie:	Ganzstahlkarosserie mit wetterfester, gebrannter Kunstharzlackierung. Seitlich große, herausnehmbare Schiebefenster. Frontscheibe aus Sicherheitsglas. Panoramascheibe. Durchsichtiges Dach, abnehmbar. Elektrischer Scheibenwischer. Verschließbarer Gepäckraum im Heck.
Innenraum:	Hintereinanderliegende schaumgummigepolsterte Sitze. Vordersitz verstellbar und hochschwenkbar, rückwärtige Sitzbank ganze Breite des Fahrzeugs. Hinterer Sitzraum vergrößert, Platz für 1 Erwachsenen und 1 Kind.
Maße:	Länge 2820 mm, Breite 1220 mm, Höhe 1200 mm, Radstand 2030 mm, Spurweite 1080 mm, Bodenfreiheit 160 mm, Eigengewicht ca. 230 kg, Nutzlast 200 kg.
Tankinhalt:	14 Liter Benzin-Öl-Gemisch, davon 2,5 Liter Reserve.
Kraftstoffnormverbrauch:	3,2 Liter/100 km.
Geschwindigkeit:	Spitze 100 km/h, autobahnfest mit 85 km/h.
Elektrische Ausrüstung	12-Volt-Lichtbatteriezündung, 90 Watt. Zwei 35-Watt-Scheinwerfer mit 105 mm Durchmesser. 2 hintereinandergeschaltete 6-Volt-Startbatterien, 12 Ah. Doppelseitige Nummernschildbeleuchtung mit Schlußlicht und Rückstrahlern, Stopplicht. Seitliches Blinklicht mit Kontrollampe

Messerschmitt-Kabinenroller KR 175

Motor	Gebläsegekühlter F. & S.-Einzylinder-Zweitaktmotor. Bohrung: 62 mm; Hub 58 mm; Zylinderinhalt 174 ccm; Verdichtung 1:6,6; Leistung: 9 PS bei 5250 U/min. Bing-Startvergaser mit Naßluftfilter und Ansauggeräuschdämpfer elektrischer 12-Volt-Anlasser
Getriebe	Vierganggetriebe mit Ratschenhandschaltung
Antrieb:	Gekapselte ⅓ × ¼″ Rollenkette auf Hinterrad.
Fahrgestell	
Rahmen:	Verwindungsfreier Stahlrohrrahmen mit vollkommen geschlossener Bodenwanne.
Lenkung:	Direkte, verschleißfeste Achsschenkellenkung mit gummigelagerten Spurstangen. Moderner Lenker mit Signalknopf.
Federung:	Einzelradfederung mit weichen Gummidruckfedern.
Bereifung:	4,00 × 8″ Spezialprofil, sämtliche Räder untereinander auswechselbar.
Bremsen:	Durch Seilzug betätigte Innenbackenbremsen. Fußbremse und feststellbare Handbremse auf alle 3 Räder wirkend.
Karosserie:	Ganzstahlkarosserie mit wetterfester, gebrannter Kunstharzlackierung. Seitlich große, herausnehmbare Schiebefenster. Frontscheibe aus Sicherheitsglas. Elektrischer Scheibenwischer. Verschließbarer Gepäckraum im Heck.
Innenraum:	Hintereinanderliegende, schaumgummigepolsterte Sitze. Vordersitz verstellbar und hochschwenkbar, Begleitsitz herausnehmbar.
Maße:	Länge 2820 mm, Breite 1220 mm, Höhe 1200 mm, Radstand 2030 mm, Spurweite 920 mm, Bodenfreiheit 160 mm, Eigengewicht ca. 210 kg, Nutzlast 150 kg.
Tankinhalt:	12 Liter Benzin-Öl-Gemisch, davon 1,5 Liter Reserve.
Kraftstoffnormverbrauch:	2,5 Liter/100 km.
Geschwindigkeit:	Spitze 90 km/h, autobahnfest mit 75 km/h.
Elektrische Ausrüstung	12-Volt-Lichtbatteriezündung, 90 Watt. Zwei 25-Watt-Scheinwerfer mit 105 mm Durchmesser. 2 hintereinandergeschaltete 6-Volt-Batterien, 12 Ah. Doppelseitige Nummernschildbeleuchtung mit Schlußlicht und Rückstrahlern, Stopplicht. Seitliches Blinklicht mit Kontrollampe

das AUTO

MOTOR UND SPORT

TEST MESSERSCHMITT-KABINENROLLER KR 200
BERICHT VON DER LONDONER MOTOR-SHOW
DER SENSATIONELLE CITROËN

HEFT 22
STUTTGART, 29. OKTOBER 1955
DM 1.20

The first experimental Davis 3-wheeler was built by Frank Kurtis. This is a plaster mock-up of a proposed 7-passenger Davis (1948).

are Three Wheels enough?

THERE ARE unmistakable signs that 3-wheeled cars are beginning an upward trend in the cycle of popularity, particularly in foreign lands where the price of a conventional new car may be equivalent to two or more years of labor at the prevailing rate, instead of only 7 months as in this country. While the cost savings of eliminating one wheel are mostly a state of mind, the fact remains that some of the lowest priced transportation available today has only 3-wheels.

The question that most people want answered is whether the elimination of one wheel is dangerous. The answer to that is: under certain circumstances a 3-wheeler may overturn where a 4-wheeler would not.

To elaborate on the "circumstances," consider the line drawing at the left. Here we have a typical 3-wheeled car with nearly equal weight distribution on each wheel, one wheel in front and two at the rear. The actual weight of this machine is unimportant, but weight distribution has been assumed to be 40/60, front/rear and the height of the center of gravity is shown as 18 inches. If we were to lift one rear wheel on

PLAN VIEW

END VIEW

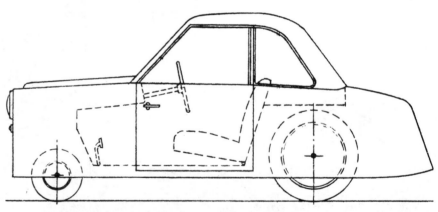

Line drawing of the AC Petite. Use of a small front steering wheel gives compact design.

The Spanish Kapiscooter is said to have a top speed of 52 mph, gives 62 mpg on "petroil"!

A home-made car built by Walter Korff.

The German Brutsch has a 200 cc two-cycle engine.

Allard's bread-and-butter model.

this hypothetical car the remaining two wheel contact points would determine a hinge-axis and this vehicle would overturn if an angle of 38° is exceeded. If two wheels with a tread of 50 inches were substituted for the single front wheel the critical angle for overturning would increase to 59°. Stated another way, the 3-wheeler would need a rear tread of 83 inches to equal the overturning angle of the 4-wheeler. However if more weight is concentrated on the rear wheels the situation improves. A fore and aft distribution of 25/75 would allow a reduction in rear tread to 67 inches (instead of 83 inches) for a turn-over angle of 59°. From this data we can deduce that a 3-wheeler should have (1) close to 75% of its total loaded weight carried by the axle and (2) tread should be close to 67% of the wheelbase, rather than the usual 50 to 55% as used on 4-wheel cars.

At the present time there are several 3-wheelers made in England, though unhappily the famous Morgan is no longer manufactured. The AC company started with a 3-wheeler (1907) and about a year ago re-entered that field with the AC Petite. Wheelbase is 72", rear tread is 42" (58.3% of wb). It weighs 840 lbs, has a 346 cc Villiers engine of about 11 bhp and sells for $765 fob. Top speed, incidently, is "nearly 40 mph". Sydney Allard has been experimenting with a 3-wheeled economy car for some time which is now in production. Here we have a color-impregnated fiberglass coupe with a 48 inch seat for 3 adults and a rumble seat suitable for two children. Power, performance and price are similar to the AC Petite, but it weighs only 672 lbs, unloaded. Mr. Allard's catalog says "a Clipper clips the cost of travel", but neglects to give wheelbase or tread.

The Regal Reliant has been manufactured in England for about two years. With a four-cylinder water-cooled engine nearly identical to the famous "old" Austin 7 it is the largest and most powerful 3-wheeler on the market. It has 16 bhp, a top speed of 60 mph, a wheelbase of 74 inches and a rear tread of 45 inches equal to 61% of

the w.b. It looks slightly similar to the Gordon and sells for about $900 fob. The Gordon is a new-comer, powered by an 8 hp, 197 cc Villiers engine. It features an electric starter and a reverse gear!

The oldest 3-wheeler on the market is the 197 cc Bond Minicar. It has been gradually improved in 6 years of production, and the

Sharp's Bond Minicar Mark C "Bear Cub". Below, the outboard engine of the Gordon.

latest Mark C version has extra seats for two children. The Bond distributor is Craven & Hendrick, Inc. 522 Fifth Ave., N.Y.C. They call it the Sharp's Bear Cub and list price is $895 in NYC.

The only serious American entry in the 3-wheel race, is the Thrif-T made by the Tri-Wheel Motor Corp., Springfield, Mass.

The chassis weighs 900 lbs, has a wheelbase of 85 inches and a tread of 42 inches. The engine is a flat twin (apparently an Onan) of 10 bhp. Price is around $800, dependent on body type. Bodies available include a pick-up, an enclosed delivery and an open utili.y model (with canvas top) which seats 5-people and luggage.

Of the German 3-wheelers only the Messerschmitt *Kabinroller* is actually in production. This strange machine is almost a motorscooter, with enclosure. A 9 bhp, 173 cc engine drives the single rear wheel in light motorcycle style. It seats two, in (tandem), but the driver steers the two front wheels by handlebar and the rear passenger is a very snug fit. Wheelbase is 80 inches, front tread is only 36.2 inches (45.2% of w.b.). Top speed is a genuine 48 mph but rapid cornering surely must be verboten. An English magazine describes this 484 lb. machine as "adequately powered and lively in performance." However, *das Auto* gives the zero to 60 kph time as 22.6 seconds (60 kph = 37.2 mph). The single rear tire (size 4.00-8) eliminates the need for a differential, but must be grossly overloaded. Summed up the Messerschmitt is a strange machine which can only be justified on the score of a very low price. It sells in Germany for DM 2375.

France builds only the "Inter" on 3-wheels, a machine slightly similar to the Messerschmitt, but with a much more stable front tread dimension, judging from photos. Spain has the "Kapiscooter" with optional roadster or delivery van bodies. There is also a choice of "Hispano-Villiers" engines; 125 or 197 cc. Wheelbase is 65 inches, rear tread is 51.2 inches, weight is 320 lbs and the price is 23.900 pesetas.

The Italians have no 3-wheelers except for certain commercial variations of their little motor scooters. However the Isetta is proving immensely popular and manufacturing licenses have been granted in France and Germany. Technically this is a 4-wheeled car, but the rear tread is only 19.7 inches. The Isetta was fully described in R & T in August 1954, page 30. ●

Willi Messerschmitt's tiny "peoples car."

German Penguin ($750) is still not in production.

U.S.-made Thrif-T Car (626 cc engine).

Your Cabin Scooter

Messerschmitt
KR 200 *DeLuxe*

in beautiful, streamlined Form!

Cheap - Fast - Manoeuverable - All round visibility - Most suitable for the winter season - Top speed: 70 mph - High performance on hills

Fully weatherproof - Excellent climber - High reselling value - Low road tax - Low insurance premium - Easy to park - Economical, running costs less than ld. per mile.

The incomparable Vehicle

The 'Messerschmitt' Cabin Scooter due to its very long development has become a very reliable vehicle which can be maintained and serviced by any one without any technical training.

Cabin Scooter De Luxe
complete with heater but without accessories.

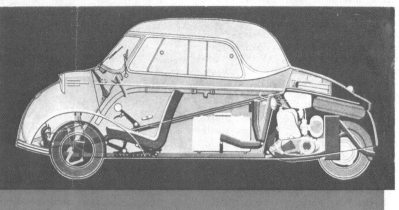

TECHNICAL DATA

Engine: SACHS 191 c.c. Two Stroke; rubber mounted; Bore 65 mm (2.5 ins) Stroke 58 mm (1.98 ins); Compression ratio 1 : 6.6, Performance 9.7 BHP 5000 rpm; Bing piston valve Carburettor with special air intake silencer and air filter; special type Exhaust silencer for exceptional quiet running; Electric starter; Fan cooling; quick acceleration; efficient hill climbing; good turning circle.

Transmission: Four speed gear box; positive stop ratchet type; silent hand operated gear change by Teleflex control system; Electrically operated Reverse gear; Four plate clutch.

Final drive: From engine by Cardenshaft to rubber mounted pinion in independantly hung Rocker box.; Roller chain fully enclosed in oil bath and driving rear wheel.

Controls: Foot pedal operated Accelerator, Clutch, Foot Brake and Dipper Switch; Hand brake with ratchet and hand grip.

Frame: Torsion proof steel tube frame with completely closed floor.

Steering: Direct steering by wear-proof divided track rods; Modern steering bar with horn button.

Suspension: Each wheel independantly suspended, with enlarged track width; soft rubber torsion springs with hydraulic shock absorbers on all three wheels.

Tyres: 4.00 × 8. with all weather steel rims and ornamental wheel discs. Spare wheel (as extra) may be placed beneath the engine cover.

Brakes: Cable activated internal expanding foot brake operating on all three wheels; Selflocking hand brake.

Body: All steel body with very durable furnace-set coat of paint; Large removable sliding side windows; Panoramic Windscreen of safety glass; Full-view Dome of organic glass; Electric windscreen wiper; Lockable luggage compartment in rear; Safety lock for cabin; Adjustable Air Deflectors, Sun Vizor and Luggage Carrier as extras.

Cockpit: Seats, one behind another (tandem) with foam rubber upholstry; Adjustable front seat can be tipped up for convenient entrance, full width back seat can be folded up; accomodation for two adults and a child.

Dimensions: Length 112 ins. Width 48 ins. Height 47 ins, Wheelbase 80 ins, Track width 43 ins, Ground clearance 6 ins, Unloaden weight 506 lbs, Pay load 440 lbs.

Performance: Maximum speed 65 mph, Climbing ability up to 1 : 3.

Electrical Equipment: 90-watt, 12 volt Lighting system; battery ignition; 2 × 35 watt headlamps; two-side illumination of number plates; twin rear lights with reflectors; Brake light; Blinkers operating in front and in rear lights; Blinker and beam indicator lights on dashboard. Ignition and reverse warning lights on dashboard.

FMR

FAHRZEUG- und MASCHINENBAU GmbH. REGENSBURG

Germany MESSERSCHMITT

MESSERSCHMITTWERKE, STUTTGART

MESSERSCHMITT

Rather odd in appearance is the three-wheeled KR-175, produced by Messerschmittwerke in Germany. Many of these tiny mini-cars are finding their way to the streets of America; they are especially attractive because of their under-$1000 price tag and their amazing fuel economy which is about 100 miles per gallon. The deluxe KR-200 is also available for 1956.

Seating is by tandem, purpose of which is to retain the center of gravity on the longitudinal axis whether one or two passengers are carried. This arrangement insures comfortable stability on curves.

The KR-175 is powered by a Fichtel and Sachs rear-mounted, air-cooled, two-cycle engine, which retains many motorcycle-like characteristics such as a rotary-type accelerator grip on the left hand side of the steering handle bar. The KR-175 engine has a displacement of 10.66 cubic inches which develops 9.5 bhp at 5250 rpm.

The vehicle has an overall body length of 111 inches and is designed to compete in the motor-scooter market with the comfort qualities of a full-scale automobile. The four-speed transmission is equipped with foot pedal clutch and is hand-operated with a shift lever at the driver's right. It has a cruising speed of over 40 mph. ■

The record-breaking Messerschmitt three-wheeler. In the successfully concluded attempts at Hockenheim the machine averaged 64 m.p.h. for 24 hours. The power unit is a 191 c.c. two-stroke

Messerschmitt Records

ON August 28 and 29 at the Hockenheim circuit in Germany, a super-sports version of the 191 c.c. KR200 Messerschmitt three-wheeler attacked a number of world's records in the 350 c.c. class, Category B (three-wheelers and cyclecars). Subject to F.I.M. confirmation, the little machine captured the 1,000-mile, 2,000-km and 24-hour records set up in 1933 by Monneret, Barthelemy and Haas on a Koehler Escoffier. The Messerschmitt's speeds (with the previous figures in brackets) were as follows: 1,000 miles, 65.8 m.p.h. (62.0); 2,000 km, 65.2 m.p.h. (61.3); 24 hours, 64.0 m.p.h. (61.5).

ROCK 'N' ROLL 'N' DRAG

BY Elvis Presley

I'VE been wild about cars almost since I can remember. When I was six years old, singing songs for folk gatherings in Tupelo, Miss., I recall feeling sad because we had to walk while all those sleek cars passed us on the road.

Yes, way back in those rough old days, when dad couldn't spare the cash to buy me a guitar, I used to dream about ridiculous things like Cadillacs and such.

As I got older, my car tastes began to change a little. I remember craving a Lincoln Continental, a classic Packard, then a Model T and finally a '32 roadster. I wanted a '32 so bad I think I'll never crave anything as much again. I dreamed of souping it up, customizing it and maybe dragging it out a little.

As I grew older — I'm 21 now — I looked more and more longingly at the antique cars I used to see in the magazines. Once, I even saw a beautiful Bugatti parked in front of a big house in Biloxi. I knew it would take money to own one of those classics. And I was sure I'd never have enough.

Cars — any kind — were out of the question for me. I was picking out tunes on a $2.98 guitar so how could I afford anything that cost $2,000? Nearest I ever got in those days was a part-time job cleaning up a garage in town.

Then, I finished high school and began to try real hard for the big time. I cut some records and almost before I knew it, I had cash. Enough of it to fix the family up comfortable and buy myself a dozen guitars. And a dozen cars, if I wanted them.

Well, I'm a little ashamed to admit it—but I bought me three little old Cadillacs, the kind I first craved 15 years ago. There's a yellow sedan, a pink convertible and a big black limousine which I can take the folks around in. Then, after I got those off my chest, I bought a three-wheeler—a Messerschmidt. This is a cool little buggy, if ever there was one, perfect for zooming around town when I'm home. And she gives me almost 50 miles to the gallon. I have a motorcycle, too, but I don't get much of a chance to ride it these days.

The most important thing, though, I still haven't forgotten. It's that '32. And one of these days—very soon as a matter of fact—I'm going to have enough time to shop for it. And as soon as things quiet down, you'll find me in the garage, chopping and channeling away like crazy. If there's any time left over,

I'll be out at the strips, maybe even competing. Rodding is for me.

Before I finish, I want to say a couple of things that need to be said. Some people make nasty remarks about hot rodders—just the way they sometimes do about rock 'n' roll music and the kids who love it. Neither are fair.

Sure, there are some irresponsible kids who break the rules. But, they're the exceptions — and that holds true about rodding as well as rock 'n' rolling. Outlaws never set the pace.

All you can ever do is your honest best and as long as it's constructive and peaceful, you can feel proud. There's little we can do about the few people who would like to destroy everything because a little part of something is bad.

DRIVING AROUND

with WALT WORON

IN MOST OF US there is an undeniable fascination for things mechanical. In a good percentage of us there is considerable interest in things of a diminutive nature. And in a large share of us there is a healthy respect for anything built by German craftsmen—as witness the astounding success of Volkswagen in this country. A study of all these points must have been made by the builders of the new tiny Messerschmitt KR-200, with the conclusion that what this country needs more than anything else is not a tax reduction, but thousands of tiny, 3-wheeled, closed scooters.

I know that some people will chide me for calling the Messerschmitt a "scooter," but I note that the factory and even many of the distributors in this country don't refer to it as a car. Instead, they say, "little vehicle," "3-wheel cabriolet," "the answer to personal transportation." I think that they are being honest in so doing, for altho it has many advantages over motor scooters, it has only the one over a car—that of providing minimal transportation at minimum cost ($1073 plus tax, in Los Angeles).

Apparently, it does the job quite well in Germany, but in this country it is more apt to find a market among those motor scooter users who would like to drive under cover in the rain, among those owners of businesses needing stimulation by promotion and publicity stunts, in rural communities where persons need drive a few short miles to and from their home,

on large film studio or manufacturing lots, on campuses, and among those who want a luxurious scooter. The best way I know of finding out what a vehicle is suited for is to drive it. With this thought in mind I approached Dan Collins of Frank Sennes Motor Corp. (210 W. Colorado, Glendale, Calif.), for a short-term loan of the "little m." For the following week I became the butt of jokesters, punsters, worried-looking head-waggers, persons seriously interested in finding out more about "that," and kids who wanted their parents to "buy me one of those."

I didn't attempt to get any performance figures, because I felt that I could get this information by pitting it against the *huge* behemoths on all sides of me in the infamous traffic of Los Angeles. Necessarily always on the alert, I would accelerate as fast as I could thru the 4 gears of the motorcycle-type transmission to keep from being run over, and steer out of harm's way with the airplane-type control bar when I couldn't stay ahead.

Everything concerned with the "little m" was a novel experience, from getting into it to pushing it hard enough thru a corner to lift the front wheel (on the outside of the curve) off the ground. To enter, you unlatch the canopy, tilt it to one side, step in and lower yourself to the driving position. The seat can be fitted to all but the very tallest, for it has 12 adjustment points on a sliding track. There is good headroom, legroom, and shoulder room. The only instrument that demands your attention is the speedometer (in kilometers per hour), which is small but easy to read. Thru some sections of the plexiglas canopy objects will distort, but vision otherwise is superb; you can see the road directly in front of your feet by leaning forward slightly. The windshield wiper is electrically operated.

Altho you would think that such a small vehicle would be exceedingly easy to drive, it's not entirely—primarily because the transmission requires a great deal of attention. Naturally, it is to be expected that you would have to go thru the 4 gears when you're increasing speed, but it's also required whenever you decrease speed or come to a stop. It takes

a fairly hard push and complete clutch pedal action to get it to stay in gear. Once you have it in gear the handle returns itself to the original position by spring tension. After you have the transmission in neutral (by use of the fingertip control on the gearshift) you pull the lever back and let it return to upright for 1st gear. Then you push it *forward* and allow it to return to upright for 2nd gear. Repeat for 3rd and 4th. To downshift, you reverse the procedure. Since there are 4 speeds in reverse also, you go thru the same procedure to back up, only *reverse* it. Sound complicated? It is!

As far as maneuverability is concerned, the "little m" really shines. You can whip it around in spots as tight as your mother's corset and park it where only scooters could thread their way in. What you don't want to do is park it where other cars can bang into you, for tho the bumpers are available (for $50 extra) they can hardly be expected to give you much protection.

Once you get onto the open road you have to anticipate road surface changes, since each one bounces you off in a different direction, with the result that you weave down the road. Due to this disconcerting action you don't feel like doing much above 35-40 mph, which is well below its claimed top speed of 62 mph. When you get close to dips and bumps it's best to slow down quite radically, for there is considerable wallowing coming out or over them. It's possible that this is due as much to the rubber torsilastic suspension units as it is to the 3-wheel setup. The double-track steering mechanism adequately insulates road shock from the steering wheel.

The passenger's seat, tho thinly padded, is wide enough to accommodate an adult, has good headroom, but leaves little legroom. It's more suited to a child or to hauling small packages. Interior finish is quite good.

You won't have to drop into your corner gas station too often since the Messerschmitt just doesn't burn much gas (anywhere from 60 mpg on up to 100, depending on how you drive it). Lubing might be a problem unless you cultivate the friendship of the station attendant, who will have to work from the owner's man-

ual the 1st few times. Repairing and maintenance shouldn't be too difficult, since the engine is a fairly simple air-cooled, single-cylinder 2-stroke, and the rest of the chassis is uncomplicated. Most motor-cycle shops would be able to remedy any difficulties if you can't get to a Messer-schmitt dealer.

Technical Data KR 200

Engine: 200 cc. fan-cooled single-cylinder F. & S. two stroke engine. Powerful acceleration, high performance on hills, higher output. Especially high engine torque with l o w r.p.m. Reverse gear to be electrically selected. Rubber mounted rear wheel suspension, which is separated from the engine. Intermediate power transmission by cardan shaft. Cylinder bore: 2,5 ins. (65 mm), Piston stroke: 1.98 ins. (58 mm), Cylinder capacity: 191 c.c., Compression ratio: 6,6 : 1, Carburettor: Bing piston valve carburettor.
Electric starter: 12 volt.

Gears, Transmission

Type: Four-speed gearbox, hand-operated; accelerator, clutch, brake and traffic beam switch are operated by footpedals. Noiseless gear shifting by Teleflex cable.

Chassis

Frame: Torsion-proof steel tube frame with completely closed floor.
Suspension: Each wheel separately suspended with enlarged track width, soft rubber torsion springs with hydraulic shock ab-sorbers on all three wheels.
Body: All-steel body with very durable furnace-set coat of paint, Large sliding side windows; Panoramic windscreen of security glass. Full-view dome. Baggage compartment in rear hood.
Cockpit: Seats with foam rubber upholstery. Front seat adjustable which can be tipped up for convenient entrance, full width back seat to fold up. Accomodation for an adult and a child.
Tank-capacity: 3 gallons petrol/oil mixture, of which $1/2$ gallon is reserve.
Performance: 10 h.p. (B.H.P.) at 5250 r.p.m.
Fuel consumption: 85—100 m. p. g.
Speed: Maximum 62 m.p.h., cruising 53 m.p.h.

U. K. Concessionaires:

CABIN SCOOTERS LTD.
17, Great Cumberland Place
Marble Arch,
LONDON W. 1

MESSERSCHMITT Cruisette

MESSERSCHMITT Cruisette

Continental styling for practical Americans

CRAIGHEAD-KOONTZ, INC.
Importers and Distributors
SOUTH BEND, INDIANA

a product of the famous
RSM MESSERSCHMITT-WERK
Regensburg (Western Zone) Germany

up to 100 miles per gallon of gasoline . . . . . priced at less than a thousand dollars !

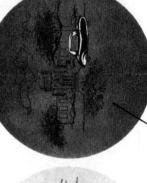

LIGHT DELIVERY
Multiple units may be purchased and operated at minimum investment and upkeep. Small area parking.

CAMPUS CAR
The safety features and economy make it ideal for the college crowd. A low priced foreign sports car.

SIGHT-SEEING
The extreme maneuverability and excellent visibility are ideal for sight-seeing trips and vacations.

SUBURBAN
The economy of operation and low initial cost are perfect for the two-car necessity in suburban living.

DAWN PATROL.—After a week's commuting, the miniature cars assemble at the "Red Lions," Bledlow, for detail tests around a country circuit and over a level measured distance.

AIR-COOLED OUTINGS

E. H. ROW and *The Motor* Staff Investigate the Possibilities of Miniature Three- and Four-wheelers for Economical Town and Local Motoring

QUITE what started it, I don't remember, but with so many readers getting hot under the collar about the economics of their day-to-day motoring in these times of financial squeeze and high living costs, it seemed a good idea to try out some of the small motorcycle-engined "mini-motors" to see whether perhaps they were an answer to the problem. The upshot was that, quite suddenly, a shoal of unfamiliar looking vehicles appeared in the Temple Press car park, to the astonishment of all concerned, and particularly to those who were to sample them.

The idea was not a new one for, back in the 'twenties, the late Ernest Perman, then general manager of Temple Press Ltd., was imbued with the same thought and caused to be built a small four-wheeled mono-car, powered by a rear-mounted 495 c.c. Scott Flying Squirrel engine. To judge from reports the vehicle left a lot to be desired and the idea was dropped. Today much more is known about making small vehicles behave properly, and great numbers have been sold post-war to graduates from motorcycling and to embryo car owners.

Most vehicles of this type are three-wheelers and, consequently, show an immediate taxation saving of £5 or £7 10s. per annum over a four-wheeler, depending on the age and size of the latter. At an annual mileage of 8,000, this amounts to .15d. or .22d. per mile. Four of the six vehicles tried out—Gordon, Bond, Messerschmitt and A.C. Petite—qualify for this

reduced taxation: the remaining two, however, a Jarc, which is a full four-wheeler, and the Isetta, which has its pair of rear wheels set fairly close together, do not.

With the exception of the Isetta, all are powered by two-stroke engines, which rely on the petroil total-loss system of lubrication. So, although they run quite happily on non-premium fuel, an oil additive in the ratio of 1 to 16 is necessary, which puts the total cost of fuel up to approximately 5s. a gallon in London—an increase of 11d. On the other hand, there is no sump draining or topping up. A further economy will be shown by the lower replacement cost of tyres; for example, the price of a complete new set of covers for a Bond amounts to approximately £9 10s.

For a week, the miniatures were driven by various members of *The Motor* staff, changing around each day, on their normal journeys to and from business and in and about London, a careful check being kept on mileage, fuel used, and the traffic conditions in which the vehicles were driven. In every case, it was found that the fuel consumption was considerably

heavier than the claims generally made, probably because the drivers endeavoured to maintain the same sort of average speed and get-away from traffic lights and hold-ups as with their normal cars: to achieve this, hard driving in the gears and wide throttle openings were the rule. Best fuel consumption under these conditions was shown by the Messerschmitt (72.5 m.p.g.) and the least good by the A.C. Petite (35.5 m.p.g.). In fairness to the latter, however, it must be recorded that 60% of its observed mileage was in dense traffic, as against the Messerschmitt's 31%. The four remaining had an average of 26% mileage in heavy traffic, and their average fuel consumption worked out at 47.6 m.p.g.

One definite advantage which these miniatures showed was an ability to use ridiculously small spaces for parking. On several occasions, quite a considerable walk was saved by being able to wheedle in where it would have been impossible to insinuate even a small normal vehicle.

One does not need to be a student of economics to realize that it is most unusual to obtain much better real value than one pays for, and, in the case of the lower-priced miniatures tried out, it must be admitted that, as passenger vehicles, they were a trifle stark. Moreover, it was necessary to have recourse to the tool kit more frequently than the average modern car driver may find acceptable. On the other hand, the two highest priced

ONE UP, ONE OFF.—The "T.P. Special" monocar built in 1928 at the instigation of the late Ernest Perman. It was powered by a 495 c.c. twin-cylinder two-stroke engine set at the rear and had quarter-elliptic springing to all wheels.

BEST WHEEL FORWARD.—On the Bond, which has f.w.d., the engine/gearbox unit is set ahead of the front wheel as may be seen in this cornering view of the vehicle.

STEPPING IT OUT.—Second fastest in the circuit tests, the A.C. Petite shows its paces and scaled-down-car lines on one of the quicker stretches.

Air-cooled Outings - - - - - Contd.

vehicles—the Isetta and the A.C. Petite—provided completely trouble-free motoring throughout, and offered quite an agreeable degree of comfort and protection.

The commuting and traffic tests concluded, the fleet (or rather shoal) was taken into the country to do several laps of a 7.6-mile circuit against the clock and each other, again with a very careful check on the fuel consumption. The course was specially chosen to show hill climbing ability, road holding, steering and braking, and included a ¾-mile climb of the order of 1 in 10, which, it was calculated, would call for prolonged use of low gear in practically every case. Assumption was borne out by fact and, amongst the smallest engined vehicles, revealed that unless one is content to make haste slowly, sparking plugs are needed which will take considerably greater punishment than those actually fitted.

The speeds recorded and fuel consumption during this part of the test are shown in Table 2; it will be seen that, with the exception of the Gordon, which was not fully run-in on delivery and became freer as the miles totted up, and the Isetta,

which has a four-speed gearbox and, anyway, is happier in open country than in town, fuel consumptions during fast driving around this hilly circuit were even higher than in London.

Thus, to judge from both the general and detailed tests, really worthwhile fuel economy from miniature motorcars can only be obtained if one is prepared to endure a slower point-to-point average than that to which the average car driver is accustomed, and drive them well within their capacity.

To obtain some idea of the speeds of which these vehicles are capable, they were finally given two runs, one in each direction, over a measured 1/10 mile. The run-in on each side was limited and in several cases, given more room to gather speed, slightly better figures than those shown in Table 3 would be obtainable: the results, however, give a fair indication of the sort of maximum to be expected from each make.

It has been previously mentioned that, in the interest of the lightness needed to obtain any sort of performance from the small power units used, there is a certain simplicity of design and equipment. To what degree this is acceptable to motorists more used to orthodoxy, and to what extent it is balanced by saving on tax and fuel, is a matter for individual consideration. Readers, therefore, are left to judge for themselves from the combined opinions of those who drove the miniature cars, and whose usual means of everyday transport consists of 1949 Vauxhall Velox, 1955 Standard Eight de luxe, 1949 Triumph 2000 Roadster, 1952 Hillman Minx, 1954 Triumph TR2, 1952 Morgan Plus 4 and 1952 Ford Zephyr.

TABLE 2

Car	CIRCUIT TESTS			
	Mileage	Best lap speed (m.p.h.)	Mean lap speed (m.p.h.)	M.p.g.
Gordon	30.4	27.4	24.3	41.9
Bond	38	30.3	27.5	41.1
Messerschmitt	30.4	32.2	30.7	33.6
Jarc	38	30.9	27.9	38.09
A.C. Petite	38	34.9	31.6	33.3
Isetta	45.6	38.00	34.8	46.77

TABL

Model (Prices include P.T.)	Capacity	Max. b.h.p.	Unladen weight as tested
Gordon ... ... (£301 2s. 9d., Tax £5)	197 c.c.	8 at 4,500 r.p.m.	6¾ cwt.
Bond Family de luxe ... (£304 7s. 8d., Tax £5)	197 c.c.	9 at 4,500 r.p.m.	5 cwt.
Messerschmitt ... ... (£309 16s. 4d., Tax £5)	174 c.c.	9 at 5,250 r.p.m.	4¼ cwt.
J.A.R.C., Mark II* (£347 16s., Tax £12 10s.)	248 c.c.	12 at 5,000 r.p.m.	6¾ cwt.
A.C. Petite de luxe ... (£392 3s. 3d., Tax £5)	353 c.c.	8.25 at 3,500 r.p.m.	8¾ cwt.
Isetta ... ... (£439 7s., Tax £12 10s.)	245 c.c.	12 at 5,800 r.p.m.	7 cwt.

* See footnote on page 140.

Gordon

The design of this vehicle is of the utmost mechanical simplicity. There is a single-tube backbone frame, carrying the individual rear wheels on radius arms, coupled to undamped coil springs. The single front wheel operates in a pair of swivelling forks, rather after the style of a motorcycle. The power unit, a 197 c.c. Villiers two-stroke, with gearbox and clutch unit attached, is located at the right-hand extremities of a pair of cross-members, and drives by chain to the offside rear wheel. The bodywork, although extremely light, is quite roomy, seating two people on a bench-type front seat (rather too upright for comfort) and having a large compartment behind for either luggage or impromptu accommodation of two children. There is a door on the left only and the weather protection takes the form of a "tourer" hood and detachable side curtains. To obtain performance the small power unit is low

THE ACCOMMODATION

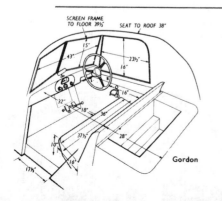

Gordon

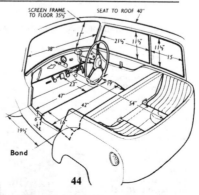

Bond

Messerschmitt

FASTEST LAP.—The unorthodox B.M.W. Isetta heels slightly as it is held through an acute bend, to record 12 minutes dead for the 7.6 miles.

"LITTLE HORSE" is the type name of the 250 c.c. Jarc, available in this country only in light van form and shown here on the long climb which was part of the country circuit.

Weight with 2 up	B.h.p. per cwt. laden	COMMUTING TESTS			
		Mileage	Comprising		M.p.g.
?½ cwt.	.82	112	Heavy traffic ... 32% Light traffic .... 14% Light traffic, 3 up 2% Open main road 61%		41.2
3 cwt.	.88	133	Heavy traffic ... 15% Light traffic ... 50% Open main road 25%		53.2
7½ cwt.	1.2	116	Heavy traffic ... 31% Light traffic ... 38% Open main road 31%		72.5
9¾ cwt.	1.2	163	Heavy traffic ... 29% Light traffic ... 20% Open main road 51%		52.0
1¾ cwt.	.7	107	Heavy traffic ... 60% Light traffic ... 5% Open main road 35%		35.5
0 cwt.	1.2	293	Heavy traffic ... 29% Light traffic ... 34% Light traffic, 3 up 1% Open main road 36%		44.2

geared and, in consequence, the exhaust note rises to a very high pitch at much above 30 m.p.h. The gear-change lever is conveniently situated in a notched gate on the driver's right-hand side and the change itself, of the quadrant variety, needs a little getting used to in order to "feel" it in. The clutch, of the motorcycle type, allows but little travel between "in" and "out"; consequently it takes time to learn how to achieve a smooth get-away. Generally speaking, the unit proved to be a good starter, but on the occasions when this was not the case the battery capacity was insufficient to allow for prolonged use of the starter motor (there is also a kick-starter). Engine accessibility, when the covering panel is removed, is quite reasonable, but it was necessary completely to remove four screws to take the panel off, and the screwdriver supplied in the tool kit was insufficiently fine to tackle the job. Slots, instead of holes in the panel securing brackets would obviate a lot of trouble.

On smooth going the ride was adequate, but over undulating surfaces the undamped coil springs at the rear produced a lot of pitch and with full load were insufficiently damped to prevent roll at speeds much above 27 m.p.h. The driver and engine being on the right, sharp left turns, as in manoeuvring quickly in a car park, were found inadvisable.

Braking at city traffic speeds was adequate, but stops from speeds above 30 m.p.h. suggested that brakes of greater efficiency would be an advantage.

Bond De Luxe

The example used was the family model which has two inward facing occasional seats behind the main bench seat to accommodate a pair of small children. As in the case of the Gordon, the bodywork with rear wings constructed of plastics material is very light, although quite roomy, and the weather protection consists of a folding canvas hood and detachable side curtains. The 197 c.c. Villiers engine, which towards the end of the tests showed a marked reluctance to start, and gearbox are mounted integral with the front wheel on a swivelling unit, and at anything approaching maximum speed the exhaust note is extremely penetrating. The fierce clutch takes some getting used to, as also does the steering column gear change which has no indication as to gear positions: particular difficulty was experienced by certain drivers in locating neutral. The Bond was found capable of keeping up with main road rush-hour traffic provided full use was made of gears and r.p.m., but a sustained 30 m.p.h. open road average proved difficult to maintain for any length of time.

The suspension is definitely firm, but road holding appeared remarkably good except for a tendency to pitch on uneven surfaces. Once one got used to its directness and strong castor action, the steering was quite satisfactory and the remarkable steering lock was considered an excellent feature, particularly when parking and manoeuvring in confined spaces; it also compensated for the lack of a reverse gear. The absence of this gear, incidentally, calls for an additional driving test by those whose driving licence does not cover Group 9 vehicles. The unit type of body-chassis construction permits of only one door being fitted—on the near side—and this, being rather small, does not contribute to ease of entrance and exit. Leg room is quite adequate for a six-footer, but with the fixed seat is too generous for drivers of smaller stature.

Messerschmitt 175

The example used was privately owned and had covered a considerable mileage before the tests. One rides rather than drives this unorthodox tandem-seated three-wheeler, it having a closer relationship to a motorcycle than to a car. A high power-to-weight ratio enables a remarkable road performance to be obtained from the 174 c.c. rear-mounted engine which

	FLYING ONE-TENTH			
Car	Best one-way time (sec.)	Best speed (m.p.h.)	Mean time (sec.)	Mean speed (m.p.h.)
Gordon ...	7.4	48.7	8.2	43.90
Bond ...	7.0	51.4	8.35	43.11
Messerschmitt ...	7.2	50	7.6	47.3
Jarc ...	6.8	52.9	8	45
A.C. Petite ...	7.3	49.4	8.1	46.75
Isetta ...	6	60	6.65	54.13

TABLE 3

THAT IS OFFERED

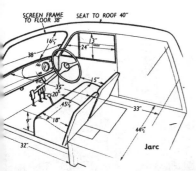

SCREEN FRAME TO FLOOR 38"   SEAT TO ROOF 40"

Jarc

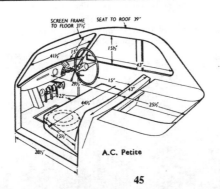

SCREEN FRAME TO FLOOR 37½"   SEAT TO ROOF 39"

A.C. Petite

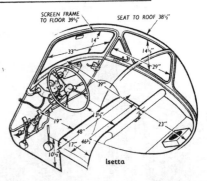

SCREEN FRAME TO FLOOR 39½"   SEAT TO ROOF 38½"

Isetta

TANDEM TWOSOME.—The unusual lines of the Messerschmitt stem largely from aero practice and the vehicle strikes a compromise between motorcycle and miniature car.

starts easily by means of a Dynastarter, but the steering and brakes on the model tested imposed limitations. The positive-stop gear shift was a decided improvement over the non-positive-stop examples, and very rapid changes could be made with ease. This vehicle also had no reverse gear and driving licence limitations, mentioned in connection with the Bond, apply.

Stiff suspension results in a rough ride on any but the smoothest surfaces, and the hinged Perspex top, while providing excellent weather protection, makes entry and exit difficult, mists up far too easily, and in some cases produced a feeling of claustrophobia.

Jarc

A true four-wheeler with Girling-damped, swing axle independent suspension to all wheels and a special Excelsior Talisman 250 c.c. twin two-stroke engine mounted under the floor slightly aft of amidships, this vehicle is in the form of a light delivery van with 3½ cwt. payload. As supplied in this country, it is provided with only a single seat, but for export markets an additional front seat is available, and provision can also be made for carrying two small children. The power unit is quiet and starts easily and silently under the impulse of a Dynastarter, is flexible and gives useful performance. The clutch is slightly less fierce than on the previous models described, and the gear change is not too difficult once one has learnt to rely upon "feel" rather than the marks on the gate. The suspension is good, particularly over bumpy surfaces, and the braking adequate: indeed, once one has become used to a notable degree of oversteer and a slight amount of roll on corners, the Jarc can be motored quite enthusiastically and with a feeling of confidence.

A.C. Petite

Of the three-wheelers, this vehicle was quite the nearest thing to a real car, with

OFFSET ENGINE, driving to one wheel and clearly seen in this photograph, is a feature of the Gordon shown on the approach to a winding, downhill section.

coachbuilt coupé bodywork having a roll-back centre section to the roof. The single cylinder 353 c.c. Villiers two-stroke engine is mounted behind the rear axle with triple Vee belt primary drive to the three-speed and reverse gear-box, final drive to the differential being by chain. Peaking at only 3,500 r.p.m., the engine has plenty of torque over a wide speed range, and it is therefore possible to proceed quite happily in mild traffic without overmuch gear changing. The gear-change itself, mounted on the steering column, is of the non-positive-stop variety and suffers from its inherent defects, but was much better than some others in this respect. The exhaust note has been kept acceptably quiet, and the Petite will cruise along happily at 40 m.p.h. under reasonable conditions. Except under fierce application the brakes are quite adequate for the job they have to do, but harshly applied they tend to lock the wheels allowing the vehicle to continue in a forward direction. A good standard of comfort is offered by the body, even for a six-footer; accommodation is limited to two people and the only space for luggage is on a big shelf behind the seat squab.

B.M.W. Isetta

This German-built Italian design is thoroughly unorthodox. Although a four-wheeler, the pair of driven wheels at the rear, without a differential, are set much closer together than those at the front; there is therefore a three-wheeled effect but no reduction of tax. The power unit,

a 245 c.c. o.h.v. model of B.M.W. manufacture needs fairly high r.p.m. to produce adequate power but is assisted by a four-speed and reverse gear-box with a change mechanism that is commendably good once one has become used to the various gear positions being in reverse to what is normal. The rather direct steering is very precise and braking is more than adequate. Entrance to the compact, almost egg-shaped body, the top of which is largely transparent plastics except for a fold-back centre panel, is through the outward hinging front, and accommodation takes the form of a bench-type seat which will comfortably hold two large or three rather small people. Relatively speaking, the engine and exhaust are fairly silent and the combination of adequate battery capacity and Dynastarter motor ensures reliable easy starting. Under most road conditions the suspension is reasonably smooth, but at certain speeds on undulating surfaces, fore and aft pitch sets in to an uncomfortable degree. The extremely compact build of the Isetta suggests great handiness in traffic, but much of the advantage is lost through an absurdly wide turning circle to the right; in fact, the vehicle shows to best advantage on suburban and open roads.

Since the above article was written, the Jarc concern has been acquired by a new company and the vehicle, which is being somewhat re-designed, will, in future, be powered by the 325 c.c. British Anzani twin two-stroke light car power unit.

TIDDLERS ON PARADE.—This line-up of five of the six "miniatures" tested shows (left to right) A.C. Petite, Bond, Isetta, Gordon and Jarc

Messerschmitt Cabrio-Limousine: Hood raisable to the right, with panorama safety glass windscreen. Foldable fabric roof with large "Zellon" rear window, complete with cover. Interchangeable with Plexiglass hood. Four sliding Plexiglass side-windows, rubber mounted and removable.

3

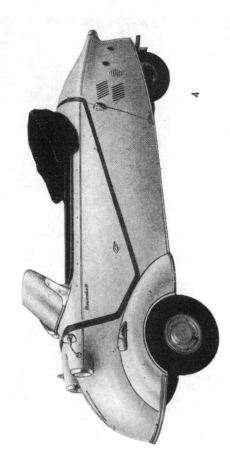

Messerschmitt Roadster KR 201: Hood raisable to the right. Fixed safety-glass panorama windscreen. Removable folding hood. Clip-on "Zellon" side windows.

4

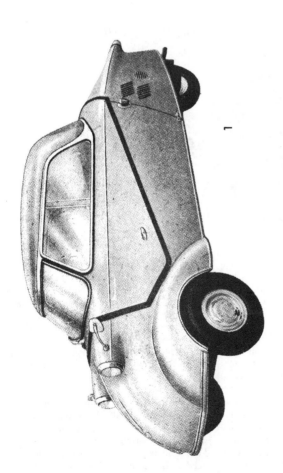

Messerschmitt Cabin Scooter KR 200: Full view hood raisable to the right, with Panorama safety-glass windscreen. Transparent removable Plexiglass roof. Interchangeable with fabric Cabrio-Limousine roof. Four sliding Plexiglass side windows, rubber mounted and removable.

1

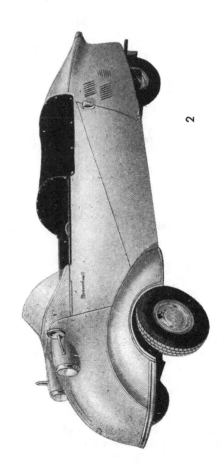

Messerschmitt Sport: Rigid body with Plexiglass sports wind shield. Removable cover extending over both seats.

2

47

U.S. Racing of three-wheelers already has begun. In the California dirt track action above, Messerschmidt drivers were pros.

Three-wheeled cars are storming in some parts of Europe. So much so that many people in the U.S. are now asking . . .

HOW ABOUT 3 WHEELS?

BY ROGER HUNTINGTON, SAE

THE THREE-WHEEL car is something like the steam car in a way . . . there will always be a few die-hards who think it's the answer to all our transportation woes. Matter of fact, we're seeing a flurry of interest in economy three-wheelers in Europe today. At least four makes are in production—Messerschmitt and Bruetsch in Germany, Isetta in Italy, and David in Spain—and they seem to be selling like hot cakes. Let's get out the slide rule and crystal ball, and see what the three-wheel car might (or might not) have to offer over here.

In the first place, I think we might note here that the three-wheel principle is being adopted only for low-priced economy cars today. Europeans aren't screaming for three-wheel *luxury* cars. This immediately suggests that we might expect to find any important advantages of this layout connected with costs.

That connection should be obvious, I think. After all, when we're thinking in terms of "minimum transportation" the three-wheeler pops up as nothing more than a natural transition stage between a *motor scooter and a conventional car!* If we think of it in these terms—and don't start dreaming of three-wheel Rolls-Royces—the concept becomes very practical. By eliminating the need for balancing the vehicle on two wheels we are able to provide regular chair-type seats, a

fully-closed body for full weather protection, and our available *space* snowballs—space either for added seats or payload. Thus, by adding one wheel we get minimum transportation that's 10 times more useful and practical than the conventional two-wheel motor scooter, and at the same time we should be able to build the thing for twice or, at most, three times the cost.

The key is the fact that we can drive just one wheel with a simple chain drive, and have the small, light-weight motor set right next to the drive wheel. The whole drive layout is strictly minimum. Four wheels are another matter. Then we need a differential, U-joints, possibly a

Three wheeler has shorter turning radius due to getting "outside" wheel closer to center of turning circle, and being able to steer it through greater turn angle.

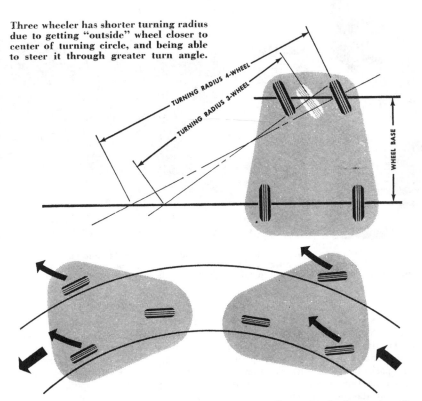

Three wheelers are either violent "under steerers" or "oversteerers" because two wheels absorb bulk of body roll force, tend to steer outside the single wheel.

In above illustration is shown the effect on a turn of having single wheel at back or front of a car. In both illustrations the vehicle is traveling toward the left.

Hot rodding of Messerschmidts is under way in Los Angeles. This one is being prepared for straightaway speed runs.

Isetta three wheeled utility car is made in Italy. Note novel door opening and the steering column attachment to the door.

Newest American three-wheeler is Basson's Star, recently announced for production, expected to sell for just under $1,000. A variety of fiberglass body styles may be offered, including utility types. Engine is one-cylinder two-cycle unit of 10 horsepower.

Among the imported three-wheelers, the Messerschmidt is best known and most popular. It seats two persons in tandem.

Three-wheeled car designed by W. E. Miller for the Arrowhead Spring Water Company was powered by a Ford V-8 engine which drove front wheels and steered through rear wheel.

Sales of Messerschmidts have grown large enough to warrant opening of dealerships, such as this one operated by Frank Sennes, in Glendale, Calif. Basic price of the car is $995.

full rear axle, etc. Our cost over the original motor scooter multiplies five or six times—but with no great increase in utility over the three-wheeler (in terms of minimum transportation). I think this is the true place of the three-wheel car in today's automotive picture.

Now when we try to expand the three-wheel concept to include larger, multi-purpose cars we get into trouble. I'm sure most of you remember the Davis car of the late '40s. Here was an attempt to market a low-priced, three-wheel "personalized" car—something on the idea of a glorified Nash Metropolitan, or maybe even a cheap four-seat Thunderbird (if there were such a thing). In other words, the Davis was intended as a car for the masses with a real sporting flavor, but a car that you could haul the family to the beach in or use to pick up the groceries. It depended for its seating space on a very wide (for that day) single "divan"-type seat, 64 inches wide. They claimed comfortable seating for four adults. Current front seat widths on full-size U.S. cars run to 65 inches, but the manufacturers don't claim four-abreast seating.

A lot of other claims were made for the Davis—through a very smart and effective publicity campaign—that threw quite a rosy glow around the three-wheel principle in those days. The unusual streamlining and light weight were to give it a top speed of 100 mph and 35-50 miles per gallon. Actually, top speed with the little 63-hp four-cylinder Continental was around 75, and a 3,000-mile economy test in 1948 showed an average figure of 28.8 mpg in overdrive. Other ballyhooed features like disk brakes, tubular frame, aluminum body, and all-coil-spring ride were abandoned on early pilot models.

Well, this is all ancient history now. The Davis project died before it ever really got rolling. I understand only some 17 pilot models were built (most of which

are still knocking around in collectors' hands). Public interest in the car seemed to run high at the time—but whether these same people would've dug into their pocketbooks, to buy the thing is another question. At any rate, it is significant that this one and only attempt to market a multi-purpose three-wheel car at a moderate price was a flop.

But let's look at the three-wheeler from a strictly engineering standpoint. Does the principle have anything to offer the American market? Probably the biggest selling point here is extreme *maneuverability*. If the single wheel is put in front your potential minimum turning radius becomes about *half* that of a normal car of equivalent size. It's a simple matter of getting the outside front wheel closer to the center of the turning circle; but an important secondary advantage is that you can turn the single front wheel through a much steeper angle without running into space difficulties with the front wheel well, as on a four-wheel car with engine and passenger compartment shoved 'way forward. The front wheel of the Davis could be turned 45 degrees, and the resulting turning radius was only 13 feet! (An average figure for a current Detroit car is 22 feet.) The advantages in city traffic and parking are obvious.

Streamlining has been mentioned as an advantage of three wheels. That is, if you put the single wheel behind, the body readily adapts to a teardrop shape. A number of interesting experimental cars have been built on this theory. Just how important is this extra degree of streamlining to road performance below 100 mph is debatable—and there definitely *is* a serious problem of aerodynamic stability with a body of this type.

Which brings up the all-important subject of handling and roadability with three wheels. It doesn't look good. The key factor here is the relative front and

rear *roll stiffness* (resistance to body roll when acted upon by centrifugal force in a curve or a crosswind). With a three-wheel car we have practically *zero* roll stiffness at the end with the single wheel. This means that the opposite end will absorb the bulk of the roll force on the body, and will tend to *steer away* from the side force. Think about this.

If we put the single wheel in front we get a violent *oversteerer*; the rear end tends to steer itself *out* of the turn, and the driver has to steer out of the turn to correct. It's touchy business. With such a high degree of oversteer a high-speed curve would be a handful for even a Nuvolari—and, in addition, an over-steering car is inherently unstable in a crosswind, requiring constant steering correction to hold it on course.

On the other hand, if we put the single wheel in back our car becomes a violent *understeerer*—that is, the front would tend to steer out of the turn, and you'd have to pull it back in. This is a more healthy situation for the amateur driver, but the three-wheeler would have such a high degree of understeer that its overall cornering power would suffer badly. In other words, two wheels would be trying to do the work of four in resisting centrifugal force in a corner. This layout would undoubtedly have a very mushy steering feel at speed in a straight line, too.

No, I guess it just won't wash. I don't see any way we could get decent high-speed roadability, tracking and handling with a three-wheel car. The principle may be great for 20-hp economy cars at speeds under 60 mph; but it's just not promising for full-size cars with 200-hp engines. Of course, we're immediately reminded at this point of the famous British Morgan three-wheel sports car that held the hearts of a small but devoted clientele for 40 years (single wheel in back). Actually, the Morgan was a

Messerschmidts also have taken to the road racing circuits. This photograph was taken at the Willow Springs course during a regular program sponsored by the California Sports Car club. Whether such road events will be repeated is unknown.

G. S. Yucker, of Los Angeles, tried this odd design in 1936. It had front-wheel drive through normal rear end assembly.

very tricky car on corners, definitely a heavy understeerer, and it required an expert driver to get any cornering out of it. Also, the car only weighed around 600 lbs. with a motorcycle engine for power; the handling problems would have been multiplied with a larger engine.

And I think that about covers it. When we get right down to cold facts it appears that the *only* important advantage that the three-wheel layout can offer in to-day's automotive picture is the possibility of a glorified *motor scooter*, with full weather protection, reasonably comfortable seating, and a good deal more seating and payload space. In view of the extremely simple drive requirements of the three-wheel layout it can easily outscore four wheels in terms of first cost. However, if we expand our field beyond this minimum transportation concept, three wheels have nothing to offer. Seating possibilities are limited, and the aforementioned handling problems are enough to kill the idea right off.

Whether the American motoring public would ever accept a three-wheel design in quantity in *any* type of car is debatable. During the Davis project a lot of people were wondering about things like a blowout on the single wheel and tip-over stability. They were skittish about riding around on three wheels. Most of their suspicions were unfounded . . . but you've always got this problem of re-educating public opinion toward any radical new design, which can cost time and money. Also, the fact that we've never really taken to a low-powered economy car here in America would make any manufacturer think twice before trying a tiny three-wheeler. (I don't think the Volkswagen can be placed in this minimum category.)

Well, you argue it out from there. Me? I'll be glad to leave the three-wheelers to the Europeans! Give me four wheels on the ground every time. ●

Gary Davis at the wheel of his first experimental car, which was built for him by Kurtis in 1940. Davis had big plans, but they evaporated before production began.

Morgan three wheeler (shot at recent Pebble Beach Concours) is one of those rare models worshipped by a cult of Morgan fanciers. This one's in top shape.

KR 200 E

MESSERSCHMITT KR 200

Der Kabinenroller will kein Auto sein,
soll auch nicht so aussehen. Auch das neue
Modell bleibt auf dem bahnbrechenden Weg
des KR 175. Denn nur durch die Sitzanordnung hintereinander
und die dadurch mögliche windschnittige Form
erreicht der Messerschmitt-Kabinenroller
seine hohe Fahrleistung und seine
Sparsamkeit im Betrieb.

Werkvertretung

KR 200 E als Kabriolett

MESSERSCHMITT KR 200

der Dreiradroller mit seinen einmaligen Vorzügen. Der Messerschmittroller wurde durch seine eigenwillige Bauart, die so viele Vorteile mit sich bringt rasch in aller Welt bekannt.

Technische Daten, die Sie interessieren.

Motor:	200 ccm gebläsegekühlter F. & S. 1 Zylinder-Zweitakt-Motor
Vergaser:	Bing-Startvergaser
Getriebe:	Viergang-Getriebe mit Handschaltung, elektr. geschalteter Rückwärtsgang.
Antrieb:	Durch Rollenkette auf Hinterrad, Kette im Ölbad
Starter:	Elektr. Anlasser
Lenkung:	Vorderräder gelenkt durch Achsschenkellenkung
Bereifung:	4,00 x 8, Platz für Reserverad unter Heckhaube
Bremsen:	Durch Seilzug betätigte Innenbackentrommelbremsen. Fuß- und feststellbare Handbremse auf alle drei Räder wirkend
Zündung:	Batteriezündung
Lichtanlage:	90 Watt, 12 Volt
Rücklicht:	Doppelseitige Nummernschildbeleuchtung und Schlußlicht, Rückstrahler, Stoplicht
Richtungsanzeiger:	Blinklicht seitlich und hinten, Kontrollampe
Karosserie:	Ganzstahl-Karosserie
Sitze:	Vordersitz hochschwenkbar und verstellbar. Hintere Sitzbank hochklappbar

Maße und Gewichte:		
	Länge	2820 mm
	Breite	1220 mm
	Höhe	1200 mm
	Radstand	2030 mm
	Spurweite	1080 mm
	Bodenfreiheit	160 mm
	Gesamtgewicht ca.	210 kg

Tankinhalt:	14 Liter Benzin-Ölgemisch, davon 2,5 Liter Reserve
Leistung:	10 PS
Kraftstoffnormverbrauch:	2,5 Liter
Geschwindigkeit:	Spitze 100 km, autobahnfest 85 km

Neuerungen, die Sie bestimmt begeistern werden.

1. Kabine: Durchsichtiges Dach, abnehmbar (Kabrio-Limousine, siehe Rückseite Prospekt), Frontscheibe gebogen (Panoramascheibe), elektr. Scheibenwischer

2. Innenraum: Vordersitz verstellbar und hochschwenkbar, rückwärtige Sitzbank ganze Breite des Fahrzeugs. Hinterer Sitzraum vergrößert, Platz für 1½ Personen auch mit geschlossenen Knien. Moderner Lenker mit Signalknopf. Fußgas, Fußbremse, Fußkupplung, Fußabblendschalter, Handbremse. Geräuschlose Schaltung durch Teleflexzug.
Geräusch in der Kabine durch hochelastische Motor-Dreipunktaufhängung besonders niedrig. Fahrgeräusch durch Spezialauspuffdämpfer und Ansaugdämpfer verringert.

3. Fahrgestell: Verbreiterte Spur
Äußerst weiche Federung mit hydraulischen Stoßdämpfern an allen drei Rädern (sehr große Federwege)
Geräuschloser Lauf der Antriebskette im Ölbad
Gummigelagerte Schwingarme
Gummigelagerte Spurstange

4. Antrieb: Größere PS-Leistung durch 200 ccm Motor, rascher Anzug, äußerst bergfreudig, erhöhte Dauerleistung.
Besonders hohes Drehmoment im **niedrigen** Drehzahlenbereich
Elektr. schaltbarer Rückwärtsgang
Gummigelagerte Hinterradschwinge, getrennt vom Motor, Kraftübertragung durch Kardanwelle

5. Karosserie: Ausgeschnittener Kotflügel
Benzineinfüllstutzen von außen

6. Ausführung: Standard-Modell wie beschrieben
Export-Modell mit folgender zusätzlicher Ausstattung
    Zweifarbig mit Zierleistenumrandung
    Blendkappen an den Scheinwerfern
    Radzierkappen
    Mit Kunstleder ausgekleideter Innenraum mit Fenster-Paneelen
    Elfenbeinfarbener Lenker
    Zeituhr

Änderungen vorbehalten.

1 KD 20

"...with the Greatest of Ease"

BRIEF ROAD TEST OF THE MESSERSCHMITT THREE-WHEELER

"NEVER mind whether it is a motor bike or a car, how does it handle in the air?" asked our resident wag when the Messerschmitt arrived at *The Autocar* offices. Un-exaggerating the question, there is indeed a strong suggestion of a light aircraft with tandem cockpit when you are sitting in this intriguing little vehicle. Even its motion and steering control have aircraft affinities. The excuse briefly to test it we regard as one of the few good things to come out of petrol rationing.

In our review of up-to-pint-sized economy vehicles, published last week, the Messerschmitt KR200, to add its rather aeronautical-sounding type number, was seen to fall in its own class between the two-wheeled scooters and miniature four-wheelers. If you have only a motor cycle driving licence you may not use its reverse; if, on the other hand, you normally drive a car, then you may use all gears in reverse as well as

forward. We are licensed to drive four-wheeled vehicles and the observations which follow are from the viewpoint of a motorist who is using this enclosed three-wheeler as an extra petrol-saving runabout for himself and members of the family. The engine is a single-cylinder, two-stroke of 191 c.c. capacity; it gives 10 h.p. at 5,250 r.p.m.

What can the Messerschmitt offer a motorist? Well, it has 2½ seats; the driver has a central one to himself and at the back there is a bench which will hold a small wife and child, or a larger wife and a shopping basket, or, again, a brother and his medium-sized dog. Behind again there is a shelf for small parcels, gloves, maps and the like. On a shopping excursion more parcels can be placed about the floor of both cockpits without getting in the way.

For those who have left their motor cycling days behind (if they have enjoyed such pleasures at all) the getting in and

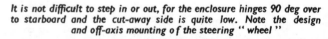

With available extras, the little vehicle can become as well equipped as a luxury car. Note the trafficators, vizor, spot lamp, screen washer and wiper. There is also a heater

It is not difficult to step in or out, for the enclosure hinges 90 deg over to starboard and the cut-away side is quite low. Note the design and off-axis mounting of the steering "wheel"

starting up are important matters. With the transparent enclosure hinged over to starboard (taking care that there is room to open it before doing so) it is easy enough to step over into the centre of the floor and to sit down. The front seat rises some six inches up and back on over-centre links to facilitate the lowering of one's person on to it. The seat must then be made to resume its normal position before starting off. The "lid" closes quite easily and latches down firmly, but it is rather heavy to lift when one wants to get out again. For support in the open position there is a strong strap which sometimes loops round the steering bar.

As for starting, one simply turns the petrol to main (or reserve); pulls out the choke, if cold; turns the ignition key to switch on (red light), and a little further to engage the electric dynamo-starter. Catch the engine on the accelerator pedal and let the choke slide in almost at once, or over-richness may stall it again. To drive backwards the engine is stopped, then the ignition key is first pressed in before going through the starting motions. This time a green ignition light comes on and the two-stroke engine runs in the reverse direction, the gears being used as for forward progress.

Economy with reasonable performance will be the desire of most owners today. The Messerschmitt cruises happily at 40 m.p.h. with two up, and it is as quick off the mark as most small cars. It covers the standing 1/4-mile in under half a minute. Driven solo, 45 m.p.h. could be used all day without pressing the engine. The brakes proved quite adequate for all purposes and speeds during the test. It is not difficult to coax 50 m.p.h. out of the machine on the level. The large vizor and other extras probably take a couple of m.p.h. off the top speed.

With hard driving—and the figure we quote was measured during a journey which included maximum speed and acceleration tests—the fuel consumption fell to 62 m.p.g. According to driving methods, 90 m.p.g. should be attainable and 75 m.p.g. might be normal for everyday use. Tyre pressures (15 lb sq in at the front, 30 lb sq in at the rear) are important to performance, stability and ride, and must be kept accurate.

The controls, particularly when steering is included, are

Two clips retain the tail, which can be hinged up, complete with fuel tank, to give access to the Sachs power unit, chain case, 12-volt battery and spare wheel

unique. The pedals, including foot dip-switch, are like those of a car. On the right is a fore-and-aft gear lever, moved by hand but operating like the foot change of a motor bike—four up, four down. A small trigger on the lever provides neutral at mid-lever position whenever it is pulled. The gear change is simple and convenient and the ratios are very happily chosen. Top gives an open-road cruising and over-drive performance; third

---

ACCELERATION: from constant speeds.
Speed Range Gear Ratio and Time in sec.

M.P.H.	Top, 4.22 to 1
10—30	.. .. .. —
20—40	.. .. .. 22.2
30—40	.. .. .. 13.2

From rest through gears to:

M.P.H.		sec.
30	.. .. ..	10.7
40	.. .. ..	20.0

Standing quarter mile 29.7 sec.

SPEEDS ON GEARS:

Gear			M.P.H. (normal and max.)	K.P.H. (normal and max.)
Top	..	(mean)	52	83.7
		(best)	53	85.3
3rd	..	..	35—44	56.3—70.8
2nd	..	..	20—30	32.2—48.3
1st	..	..	12—15	19.3—24.1

Acceleration figures are the means of several runs in opposite directions.

SPEEDOMETER CORRECTION: M.P.H.

Car speedometer	6	18	29	41	52	54
True speed	10	20	30	40	50	52

Measurements in these 1/8 in to 1ft scale body diagrams are taken with the driving seat in the central position of fore and aft adjustment and with the seat cushions uncompressed

SPECIFICATION

ENGINE: Single-cylinder Sach's, air cooled
Bore and stroke: 65×58 mm (2.56×2.28in).
Displacement (c.c. capacity): 191 c.c.
Valve position: ported two-stroke.
Compression ratio: 6.3.
Max. b.h.p. (gross): 10 at 5,250 r.p.m.
Battery: 12 volt.

TRANSMISSION
Clutch: 4-disc plate. No. of speeds: 4.
Overall ratios: Top: 4.22, 3rd: 6.06, 2nd: 9.05
1st: 17.70, Rev.: Ratios as above. Final drive: chain to rear wheel.

CHASSIS
Brakes: F & R: drum, cable operated.
Suspension: F & R: rubber torsion springs.
Shock absorbers (type): F, telescopic, double-acting, hydraulic.
Steering (type): direct linkage, no steering box.

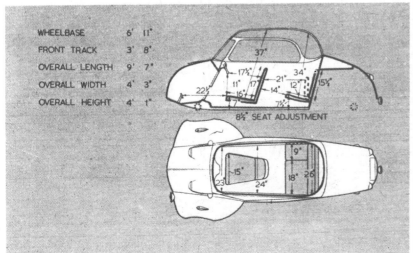

WHEELBASE	6' 11"
FRONT TRACK	3' 8"
OVERALL LENGTH	9' 7"
OVERALL WIDTH	4' 3"
OVERALL HEIGHT	4' 1"

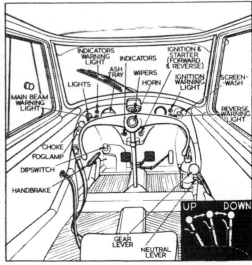

A good idea of the comparative size of the Messerschmitt is given by this photograph in Windsor Great Park. The overall width is 4ft, and that of the Vanguard 5ft 9in. An Austin A35 is 4ft 7½in wide

"... with the Greatest of Ease"

is a very flexible gear for flowing traffic and it looks after the normal speed bracket 18-40 m.p.h. Second gear suits denser traffic from zero (not too heavy a load) to 25 m.p.h.

The steering, by a wheel-cum-bars control, calls for a small, two-dimensional movement. It is a bike/car compromise. A parking brake of rather unattractive design is located beside the driver's left calf.

What of comfort and control? The suspension is firm but by no means hard. With little wheels and a short wheelbase it is only reasonable to take rough roads and pot holes slowly. Few main roads would call for less than 30 m.p.h. and the degree of bounce and pitch seems roughly proportional to speed. On normally good surfaces the ride is pleasant at all speeds and, provided tyres and suspension are in good shape, the machine controls well. If tyre pressures are uneven, lateral and directional stability deteriorate. The ride sensations of the driver are slightly accentuated for the passenger.

Until the driver is familiar with the feel of the ride and the unconventional steering, he may suffer from lack of confidence and believe the Messerschmitt to be likely to roll over on corners. In fact it would need very clumsy handling to lift a wheel or lose control when turning. The machine is best driven round a corner rather than trailed or free-wheeled round it.

At speed a quite small deflection of the steering bars can produce an abrupt swerve, but a fairly heavy load is required to move the control when travelling fast or slow. The hands may be moved closer to the column for guiding the vehicle when driving at speed on the open road. The lock is poor, and

often it would be easier to park in an awkward place by getting out and pushing than by stopping and starting, first forward and then backward. It is fair to say that the steering needs getting used to, but is then accepted without worry or complaint and can be used to give precise driving.

The clutch is quite smooth, and take-off with full load up a steep gradient gives no trouble at all. In traffic one may drive a little daringly when in a hurry and, using the gears freely, weave and queue-jump with the motor bikes. Car and lorry drivers generally treat you with amused tolerance if they see you—but that is a big "if." So small and low is the Messerschmitt that the driver must make full allowances for not being seen by others in full-sized vehicles on his right. A small mast with pennant might be fitted to mark one's position in traffic.

All sorts of fittings adorned the Messerschmitt tested; some were extras. The screen washer and the wiper are very useful because in dirty weather the little machine is likely to get well doused by the rear wheels of the vehicle ahead.

We had several outings in heavy rain, but on no occasion were we leaked upon—a creditable effort by the makers. With the heater working on damp days the condensation is rather annoying, but air from the stiff-to-operate sliding side windows or a leather with which to wipe around, get over the trouble after a time. An average-sized man can wear his hat and arrive at his office in the Messerschmitt, warm, dry and tidy. So can his passenger, but tandem seating always seems unfriendly.

For long journeys rather more seating comfort would be appreciated; the rubber matting and leathercloth trim provide both draught and sound proofing. At no time is the engine noisy or obtrusive; rather one notes its smooth willingness and flexibility.

Since the model was first introduced there have been substantial improvements all round, and now the KR200 is a little vehicle which most drivers would be pleased to own. There always seems to be room to park it and, in spite of its small proportions, it is neither cramped nor fragile.

It is estimated that there are roughly 350 Messerschmitts in this country at present and small batches keep coming in. At the moment there may be a short waiting list for prospective new owners. The spares position is said to be fairly good and is improving. Reports from Germany suggest that production of this model may soon cease, but enough essential frame and body spares such as enclosures and wings will be available. The Sachs engine, which has several other applications, remains in production. The vehicle tested was provided by M.P.H.W. Sales, of 7 Station Approach, West Byfleet, Surrey. This is also the official address of a new Messerschmitt Owners' Club, the secretary of which is Mr. "Mike" Morris.

Messerschmitt
KR 200 DeLuxe

in beautiful, streamlined Form!

Cheap - Fast - Manoeuverable - All round visibility - Most suitable for the winter season - Top speed: 70 mph - High performance on hills

Fully weatherproof - Excellent climber - High reselling value - Low road tax - Low insurance premium - Easy to park - Economical, running costs less than 1d. per mile.

MESSERSCHMITT
KR 201
SPORTS CONVERTIBLE

Sporting ■ Fast ■ Economical ■ Reliable ■ Highly manoeuvrable

TECHNICAL DATA

Engine: SACHS 191 cc. Two Stroke, rubber mounted; Bore 65 mm (2.5 ins) Stroke 58 mm (1.98 ins); Compression ratio 1 : 6.6; Performance 9.7 BHP 5000 rpm. Bing piston valve Carburettor with special air intake silencer and air filter, special type Exhaust silencer for exceptional quiet running; Electric starter; Fan cooling; quick acceleration, efficient hill climbing; good turning circle.

Transmission: Four speed gear box; positive stop ratchet type; silent hand operated gear change by TeleFlex control system; Electrically operated Reverse gear; Four plate clutch.

Final drive: From engine by Cardenshaft to rubber mounted pinion in independently hung Rocker box.; Roller chain fully enclosed in oil bath and driving rear wheel.

Controls: Foot pedal operated Accelerator, Clutch, Foot Brake and Dipper Switch, Hand brake with ratchet and hand grip.

Frame: Torsion proof steel tube frame with completely closed floor.

Steering: Direct steering by wear-proof divided track rods; Modern steering bar with horn button.

Suspension: Each wheel independantly suspended, with enlarged track width, soft rubber torsion springs with hydraulic shock absorbers on all three wheels.

Tyres: 4.00×8. with all weather steel rims and ornamental wheel discs. Spare wheel (as extra) may be placed beneath the engine cover.

Brakes: Cable activated internal expanding foot brake operating on all three wheels; Selflocking hand brake.

Body: All steel body with very durable furnace-set coat of paint; Large removable sliding side windows; Panoramic Windscreen of safety glass; Full view Dome of organic glass; Electric windscreen wiper; Lockable luggage compartment in rear.; Safety lock for cabin; Adjustable Air Deflectors, Sun Vizor and Luggage Carrier as extras.

Cockpit: Seats, one behind another (tandem) with foam rubber upholstry; Adjustable front seat can be tipped up for convenient entrance; full width back seat can be folded up; accomodation for two adults and a child.

Dimensions: Length 112 ins, Width 48 ins, Height 47 ins, Wheelbase 80 ins, Track width 43 ins, Ground clearance 6 ins, Unloaden weight 506 lbs, Pay load 440 lbs.

Performance: Maximum speed 65 mph. Climbing ability up to 1 : 3.

Electrical Equipment: 90-watt, 12 volt Lighting system; battery ignition; 2×35 watt headlamps; two-side illumination of number plates; twin rear lights with reflectors; Brake light, Blinkers operating in front and in rear lights; Blinker and beam indicator lights on dashboard. Ignition and reverse warning lights on dashboard.

CONVERTIBLE with De Luxe finish. Handsome black interior upholstery picked out in imitation snake skin. New Rear View Mirrors, Chromium Plated Rear Lights. Additional Air Inlet for Engine. Hood Cover in Black with silver piping.

SPORTS CONVERTIBLE

MESSERSCHMITT
KR 201

Sporting ▪ Fast ▪ Economical ▪ Reliable ▪ Highly manoeuvrable

CONVERTIBLE with De Luxe finish. Handsome black interior upholstery picked out in imitation snake skin. New Rear View Mirrors. Chromium Plated Rear Lights. Additional Air Inlet for Engine. Hood Cover in Black with silver piping.

 **FAHRZEUG- UND MASCHINENBAU GMBH. REGENSBURG**

SOLE CONCESSIONAIRES: CABIN SCOOTERS (ASSEMBLIES) LTD
80 GEORGE STREET, LONDON. W.I. - TEL. HUNTER 0609

Technical Data KR 201

Engine: 200 cc. fan cooled single-cylinder F. & S. two stroke engine. Powerful acceleration, high performance on hills, high output. Especially high engine torque with low r. p. m. Reverse gear electrically selected. Rubber mounted rear wheel suspension, which is separate from the engine. Intermediate power transmission by cardan shaft. Cylinder bore: 2,5 ins. (65 mm), Piston stroke: 1,98 ins. (58 mm), Cylinder capacity: 191 c. c., Compression ratio: 6,6:1, Carburettor: Bing piston valve carburettor.
Electric starter: 12 volt.

Gears, Transmission

Type: Four-speed gearbox, hand-operated; accelerator, clutch, brake and traffic beam switch are operated by footpedals. Noiseless gear shifting by Teleflex cable.

Chassis

Frame: Torsion-proof steel tube frame with completely closed floor.

Steering mechanism: Front wheels steered by divided track rods, rubber mounted. Modern steering bar with horn button.

Suspension: Each wheel separately suspended with enlarged track width, soft rubber torsion springs with hydraulic shock absorbers on all three wheels.

Body: All-steel body with very durable furnace-set coat of paint, plyable side curtains; Panoramic windscreen of safety glass. Convertible top (cabrio-limousine). Electric wind screen wiper. Baggage compartment in rear hood.

Brakes: Internal expanding brakes activated by cables, foot brake and self-locking handbrake acting on all 3 wheels.

Cockpit: Seats (one behind another) with foam rubber upholstery. Front seat adjustable which can be tipped up for convenient entrance, full width back seat to fold up.

Electrical Equipment: 12 volt, 90 watt Dynamo. Coil Ignition. $^2/_{35}$ watt Headlights. Flasher Lights in Side and Rear Lights. Parking Lights on Wings. Starter operated by Ignition Key.

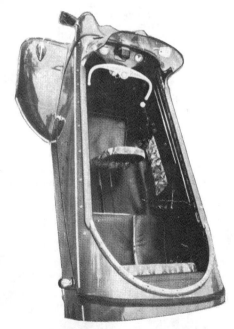

The "MESSERSCHMITT" Cabin Scooter has achieved such a perfection throughout the years, that anybody without any technical knowledge can drive and service the "MESSERSCHMITT".

Specification may be changed without notice!

FAHRZEUG- UND MASCHINENBAU GMBH REGENSBURG

Regensburg, Lilienthalstr. · Fernruf 3 07 73/3 07 74 · Fernschreiber 06 5882 · Gleisanschluß: Rgbg.-Hbf., Anschluß Messerschmitt
DISTRIBUTOR

Is this the Messerschmitt Limo !!!

TANDEM FOR TWO

ROAD IMPRESSIONS OF THE DROP-HEAD MESSERSCHMITT

WHEN is a bubble car not a bubble car? Presumably the question is answered by a modified version of the drop-head KR 201 Messerschmitt cabin scooter, which it has now been possible to try over an extensive mileage. The model's limitations—some of them inherent in the basic design—are several, yet the overall impressions are favourable. In many ways the little vehicle provides pleasure, often mixed with amusement, and it covers the ground remarkably quickly with outstanding economy.

A description of the enclosed car, complete with some acceleration figures, specification and scale drawings, appeared in *The Autocar* of 11 January last. In its latest form the car retains the 191 c.c. Sachs single-cylinder, air-cooled, two-stroke engine, running in the modified form on a compression ratio of 8.0 to 1, compared with the 6.3 to 1 of the 200 and standard 201, and giving perhaps 12 instead of 10 gross b.h.p. at 5,250 r.p.m. It is mounted at the rear, and drives the single back wheel. For reverse, a control permits the engine to be run in the

Imitation reptile skin is used in the drop-head model. The pedals are like those of an orthodox car and the gear lever, on the right, has a ratchet with motor cycle type action

This modified 201 has a prominent sports silencer protruding at the rear. The hood folds away neatly, with a smarter appearance than that suggested in this illustration

opposite direction of rotation, and as the turning circle is not particularly good, manœuvring can sometimes involve the fussiness of starting and stopping the engine a number of times. However, in many circumstances the car may be simply wheeled to and fro into its parking position.

Cabin Scooters Assemblies, Ltd., 80, George Street, London, W.1, provided the car complete with the sports silencer. This protrudes at the rear to an extent which makes some care necessary when walking round the back of the car. The noise level is also affected, being quite considerable in the sports car fashion. The 200 originally tested reached 40 m.p.h. in 20sec, while the modified 201 reached 50 m.p.h. in only 22.1sec. Figures for the standing start quarter-mile were 29.7sec and 25.1sec respectively. There was no loss of flexibility, and on top gear this 201 drop-head went from 20 to 40 m.p.h. in 18.7sec, compared with 22sec for the standard 200 tested previously.

On one run on the flat with a following breeze, the little car reached 60 m.p.h.

The plastic sidescreens, here displayed on the rear seat, are normally stowed in the compartment for oddments above the engine. The driving seat lifts up and back to facilitate entry

Thus the all-round performance is very similar to that of the average small family saloon. The 8in wheels and light weight of the Messerschmitt provide a ride in which road surface irregularities are passed to the occupants, but, on the average, smooth British main road the driver and passenger are not jolted about to any serious extent. However, there is sometimes a slight weaving at full load which is more a cause for apprehension on the part of passenger than driver. This, coupled with the direct steering (no steering box) calls for some skill in the matter of sustaining high speeds with security.

The drop-head modification of the coachwork has been accomplished in a commendably neat way. In effect, the "bubble" of the closed car is removed, and instead there is a hood, complete with flexible plastic sidescreens. The hood is easy to erect and fold, and when down is quickly and tidily covered with well-tailored matching fabric. The screens are easily fitted, but when in position they preclude hand signalling—or even disposal of cigarette ends. (A clock has taken the place of the ashtray.) Another difficulty is that it is not possible to adjust the driving mirror which is mounted externally. However, the hood and side screens provide a snug interior, free from draughts and leaks.

Throughout the test the Messerschmitt was driven hard, with many London miles included, yet the minimum m.p.g. measured was 62—no worse than that obtained with the enclosed car with the standard engine. In normal driving, cruising in the 30s and 40s, a very much higher m.p.g. would be obtainable. The engine started easily, hot or cold, it never faltered, and the tickover remained at its ideal setting. All the controls operated in response to light pressure. They are like those of a normal car with the exception of the fore and aft, ratchet gear change of the motor cycle type.

This scooter appeals as a second vehicle for domestic use, yet it is also a serious means of brisk, good-road transport for those for whom the amount of accommodation available suffices.

MESSERSCHMIDT (German)

Messerschmidt was a successful airplane manufacturer. When he began to build cars, he still made them look and act like airplanes. As a result frustrated pilots "fly" around Germany in three-wheel Messerschmidts, driving as though they were flying low. A four-wheel version of the tandem-seated car, the Tiger, has more power and "flies" even lower!

New for 1959 is a four-wheeled bench-seated coupe or roadster (above) that is said to be licensed for manufacture in the United States. It has a rear engine of about 600 cubic centimeters. The entire rear of the body lifts up to uncover the power plant.

Price is expected to be about $900.

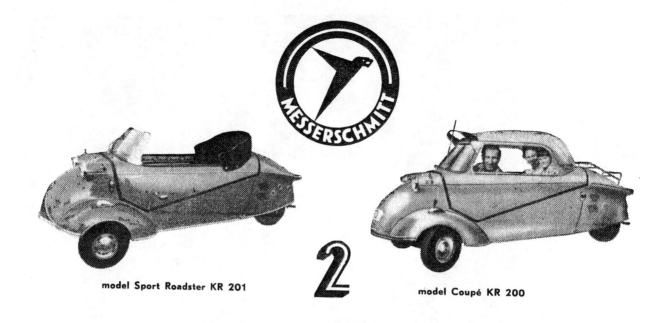

model Sport Roadster KR 201   model Coupé KR 200

VERGELIJK DE CIJFERS EN . . .

Ook U zult dan direct begrijpen, waarom er momenteel vele weggebruikers overtuigde MESSERSCHMITT berijders worden, want . . . .

Een middenklasse Automobiel kost aan :

Benzine 1 L. op ± 12 KM	Een MESSERSCHMITT 1 L. op ± 30 KM, besparing dus 150%
Een motorrevisie ± f 450,—	Een MESSERSCHMITT ± f 150,— , besparing dus 66%
Een W.A. verzekering ± f 110,— p.j.	Een MESSERSCHMITT f 42,50 p.j. , besparing dus 64%
Wegenbelasting ± f 80,— p.j.	Een MESSERSCHMITT f 30,— p.j. , besparing dus 63%
Een Buitenband ± f 74,— p. stuk	Een MESSERSCHMITT ± f 37,— , besparing dus 50%
Stalling ± f 25,— p. maand	Een MESSERSCHMITT ± f 15,— p.m. , besparing dus 40%

èn . . . . . voor *ACTIEVE MENSEN* minstens zo *BELANGRIJK* :

door handige manoevreerbaarheid in stadsverkeer én door het veel vlugger vinden van voldoende parkeerruimte :

AAN TIJDSBESPARING MINSTENS 50 % !

MESSERSCHMITT RIJDEN BETEKENT TEVENS: .

* ⋆ een topsnelheid van 100 KM per uur,
* ⋆ een vlotte kruissnelheid van 80 KM per uur,
* ⋆ een klimvermogen tot 33%.
* ⋆ een regelbare verwarming mèt defroster bij de lage prijs inbegrepen,
* ⋆ een panoramische voorruit met een ongeëvenaard vrij uitzicht naar alle kanten,
* ⋆ een veiligheidsdak zónder bijbetaling,
* ⋆ 4 versnellingen (4e als „overdrive") èn achteruit,
* ⋆ zeer snel optrekken,
* ⋆ een laag bij de weg liggend zwaartepunt, dus ongeëvenaard veilig bochtenwerk,
* ⋆ zitplaatsen op de lengte-as, een volmaakte in „balans" ligging dus, daardoor géén „duiken" in de bochten,
* ⋆ **EEN RIJBEWIJS voor Motor (of Scooter) reeds voldoende !**

MESSERSCHMITT Import voor Ned. :

Zelfs één proefrit zal ook U enthousiast maken en deze wordt U GRATIS aangeboden door :

Maassluis, Telefoon 01899-2452

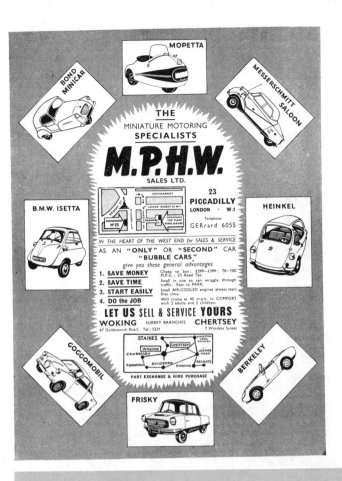

European Newsletter

And now, swinging from the sublime to a car that is anything but ridiculous (though it does look a little that way at first), we have received an enthusiastic story from Vic Hyde, a one-man band who succumbed to the lure of the little Messerschmitt (is it a car? A scooter?) while touring Europe. He writes: "It has three wheels, tire size four by eight, the single wheel in the rear. The engine is a one-cylinder two-stroke, air-cooled, mounted in rear. Has foot clutch (some are automatic), four speeds ahead, reverse, six-volt battery in two sections, radio with headphones. I have driven my Messerschmitt 6000 miles. What a thrill to drive! Not in a big-car sense, but in the feeling of sitting in a plane cockpit, with the bark and whine of the little engine in back, and the weaving in and around slow traffic. I can't afford a private small plane, but I sure can make my dreams realistic!"

Honored (for engineering) name of Messerschmitt adorns minuscule tandem-seated car of one-man band Vic Hyde

Tiny rear-mounted gas tank thriftily dispenses a gallon of fuel for every 45-100 miles of driving. Car has rear engine

With its plastic bubble in place, Messerschmitt holds two dry passengers on a rainy day, but cooks them when it's hot

The earth hath bubbles* . . .

One week with a Messerschmitt cabin scooter

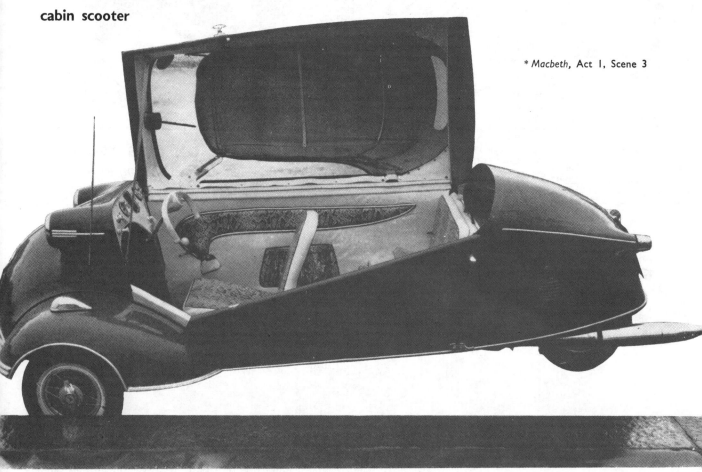

* *Macbeth*, Act I, Scene 3

ALTHOUGH they may be anathema to the dyed-in-the-wool, four-feet-from-the-ground vintage enthusiast, the number of " bubble cars " on the road is increasing steadily and we felt that the time had come to get more closely acquainted with one of these strange devices, so that *Autocourse* readers could judge for themselves the value of such miniatures as a means of transport. With the co-operation of Alan Betts Associates, Ltd., we have therefore been wearing a Messerschmitt cabin scooter loosely round the hips for a week, and now find that our somewhat suspicious attitude towards them has been very considerably modified. In the course of the week's trial we commuted daily between home and office, covered 120 exceedingly wet miles in following the Brighton Veteran Run, and even used the little Messerschmitt to attend the B.A.R.C. annual dinner dance (complete with medium-sized wife in evening dress).

Direct steering

The Messerschmitt cabin scooter is unusual in employing tandem seating, and therefore bears less resemblance to a goldfish bowl than its competitors. At the front of the sturdy tubular steel frame there is direct steering by divided track rods to the two front wheels, and at the rear a single wheel is chain-driven from the four-speed gearbox, which in turn is shaft-driven by a 191 c.c. Fichtel and Sachs two-stroke engine. The springing medium at all three wheels is bonded rubber in torsion, and hydraulic dampers are fitted all round. The brakes are mechanically operated through cables. One of the first of the miniatures, the Messerschmitt has been much modified in detail during the four or more years of its existence. Handlebar steering is still used, but of improved shape, and a foot throttle is fitted instead of a twistgrip. The engine capacity is now 191 instead of 174 c.c. and, by allowing the engine to run backwards, a Siba dynastart now provides four reverse gears instead of none at all.

Instead of the usual transparent plastic top, the model which we tried was fitted with a folding canvas hood, and the compression ratio of the engine had been raised from its standard 6·3 to 1 to an impressive 8 to 1, providing 12 b.h.p. instead of 10 at 5250 r.p.m. With hood down, the open Messerschmitt proved a delightfully " sports car " conveyance, but it lost its major advantage of all-round visibility when the hood was erected.

Getting into the Messerschmitt is not difficult, for the top is hinged at the right-hand side and the seat lifts up and back on a special linkage, but we found it even simpler to hop over the side when the hood was down. The driver's seat, functional rather than luxurious, is adjustable over an extremely wide range and ample leg-room is available. The rear seat is designed to accommodate an adult and child, but the editorial wife and eight-year-old felt a little cramped together. Behind the rear seat is a small compartment which will take a spare wheel, the flexible plastic sidescreens, or small pieces of luggage—but not simultaneously! When the top is locked in the closed position, this compartment is inaccessible and therefore thief-proof.

The pedals are conventional, and the right-hand gearchange is identical to that of a 500 c.c. racing car; that is to say, one moves it forward to change down and back to change up, the lever being returned to a central position by means of a spring, and the gearchange is very rapid indeed. A small additional finger control allows neutral to be selected between any two gears.

Stalling in Park Lane

Our first acquaintance with the Messerschmitt was not made under ideal conditions; in the centre of London at the tail-end of rush hour, after dark, in heavy rain. Attempts to start the engine proving successful after remembering to turn on the petrol tap, we set off and stalled no less than four times in Park Lane, for the clutch is on the fierce side. By far the most disturbing feature, however, was the steering, and our progress was extremely erratic for the first few miles. The combination of absolutely direct steering and a handlebar arrangement instead of a wheel can be quite frightening; try to steer the Messerschmitt like an ordinary car, and it will leap over the nearest hedge. Eventually we adopted the old dodge of the solo motorcyclist turned sidecar driver—keeping only one hand on the handlebar. This worked perfectly, and soon we were sufficiently familiar with its behaviour to steer this cabin scooter as it should be steered. As a colleague put it after a trial run, " You don't so much steer as turn your head from side to side to go round corners!" Once the unconventional steering *has* been mastered, the most remarkable manoeuvres become possible and one's cornering speed appears to be limited by timidity rather than centrifugal force. The Messerschmitt being beautifully balanced, one's motoring experience may be enriched on wet roads with the rare joy of the three-wheeled drift.

Driving the Messerschmitt, it is scarcely possible to credit the fact that the engine capacity is less than 200 c.c. The acceleration, aided by the speedy gearchange, is really excellent, and there is a great deal of mischievous fun to be had when starting from traffic lights in company with larger vehicles. Because of the tandem seating layout the frontal area is comparatively small, so that even with only 12 b.h.p. quite a high top speed is attainable. We did not reach the 75 m.p.h. which is claimed for the high-compression version, but several times exceeded 65 without difficulty. Our maximum speed tests were spoiled by stormy weather, for in strong winds the conviction grows on the Messerschmitt driver that he is about to become a Messerschmitt pilot.

Nor did we achieve 85 to 100 m.p.g., a careful test returning a rather disappointing 52 m.p.g. This included several runs with a passenger and a good proportion of hard driving, when small two-strokes do tend to become thirsty. Another point of criticism concerned the suspension, which is definitely on the hard side. Although the ride is comfortable on good surfaces, we found one bumpy road on which the Messerschmitt pitched so much as to become quite uncontrollable, suggesting tired shock absorbers on the example we tested.

Ideal in traffic

On the other hand, we covered some 400 miles at average speeds more usually associated with sports cars, in miserable weather conditions, yet always remaining snug and dry. At night we were often dazzled by dipped headlamps because of the Messerschmitt's low build, but daytime driving was unadulterated pleasure, especially when dry weather allowed us to dispense with the hood and the last vestige of claustrophobia. In heavy London traffic the little cabin scooter really came into its own by virtue of its small width, sparkling acceleration and quick steering, while all the usual parking worries could be conveniently forgotten. Above all, the Messerschmitt is *fun* to drive, so that in seven days we developed an odd affection for this fast-moving three-wheeler and felt rather regretful when the time came to return it to its owners.

The Messerschmitt in its normal plastic-top form, showing the disposition of the seats, controls and engine

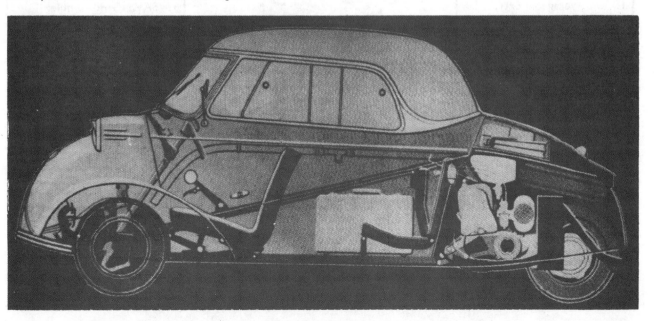

Like the three-wheeled Messerschmitt, the TG 500 four-wheeler has tandem seating, with entry through a side-hinged plastic dome.   The engine is enclosed by a quickly removable cowling

Four-wheeled Messerschmitt

FURTHER details are now available of the new Messerschmitt TG 500 which, as announced in last week's issue, is obtainable in this country. Construction closely follows that of the existing three-wheeled model, but the engine is a 490 c.c. twin-cylinder two-stroke, air-cooled by a fan for each cylinder. Bore and stroke are 67×70 mm, compression ratio is 6.5 to 1, and claimed power is 20 b.h.p. at 5,000 r.p.m. Transmission, in unit with the engine, consists of a two-plate clutch, four-speed and reverse gear box, and spur gear and differential final drive. Rear suspension is independent, by single wishbone and coil spring with concentric telescopic dampers.

At the front, springing is by rubber in torsion, and the chassis is of welded tubular construction. A $6\frac{1}{2}$-gallon fuel tank with $\frac{3}{4}$-gallon reserve is mounted between the engine and the passenger's seat, which is immediately behind the driver. Ignition, charging and starting is by 12-volt Bosch Dyna-Starter, and controls for head lamps, indicators and horn are on the steering bar.

Performance of the new model should be considerably higher than that of the three-wheeler version, and a fuel consumption of 52 m.p.g. is claimed. The price in this country is £435 basic, £651 5s 8d including purchase tax. Leading dimensions are: wheelbase 6ft $2\frac{1}{2}$in, front track 3ft 8in, rear track 3ft 5in, overall length 10ft, and height 4ft 2in. Dry weight is 770 lb. Cabin Scooters (Assemblies), Ltd., 80, George Street, London, W.1, are sole concessionaires for the United Kingdom.

The 191 c.c. Two-stroke
MESSERSCHMITT KR201

A Convertible Three-wheeled Cabin Scooter with Astonishing Performance

"On dry days, the weathertight, fold-back hood was opened . . . and the KR201 used just like any four-wheeled sports car."

IT is very hard to place the 191 c.c. two-stroke Messerschmitt *Kabinenroller* three-wheeler into a definite vehicle class. Is it a light car? Or a luxury three-wheeler? Perhaps, as *Motor Cycling*'s Road Test of December 9, 1954, suggested, it is best described as an enclosed scooter with automatic stability imparted by the use of a third wheel.

But whatever one's approach to the KR201 it must include a large measure of fascination and admiration for this extraordinary small vehicle, with all its car capabilities. Thanks to its four-speed gearbox and reversing engine, complete enclosure almost to car specification, and a most exhilarating performance achieved without excessive fuel consumption, the latest *Kabinenroller* to be imported by Cabin Scooters (Assemblies), Ltd., must be treated with respect in spite of its diminutive size.

It is, of course, the excellent aerodynamic properties that stem from the small frontal area of the attractively shaped body that go to produce a true maximum speed in excess of a mile a minute—and almost 50 m.p.h. in third. Furthermore, this maximum is allied to a usable continuous cruising speed in the upper fifties, and is achieved with a two-stroke engine of just under 200 c.c. By using the four-speed positive-stop hand-change gearbox and the peppy Sachs, fan-cooled, engine to the full, the pilot could extract such performance that the expression of surprise on the overtaken driver's face alone made the inclusive £340 purchase price feel fully worth while!

Full throttle could be indulged in *ad lib* at a commendably low noise level and, with the high-compression conversion kit fitted to the test machine, gave a *minimum* economy as good as 60 m.p.g. With more sedate driving the consumption fell to a remarkable 85 m.p.g., a petroil ratio of 24 : 1 being used. The driver could not be caught napping by his fuel level as an adequate reserve is provided.

On dry days the weathertight, fold-back hood was opened with advantage and the KR201 used just like any four-wheeled sports car. This statement is made quite without qualification, for the 500-lb. three-wheeler could more than hold its own on acceleration, cornering and braking with many bigger and more powerful vehicles. It could be flung into bends at a high knottage, the driver secure in the knowledge that no

danger existed of it turning turtle, lifting a wheel or starting to spin. The tyres screamed their protest, it is true, but they retained their grip and enabled dry drifts to be executed in the best " four-wheeled " tradition. On wet roads the rear end broke away first—a safe state of affairs. Even when in the " seventies "—possible on downhill stretches—there was no sign of instability.

Reasonable comfort was obtained from the suspension, either one- or two-up, and the brakes were sufficiently powerful to evoke tyre squeal; no fade was experienced, even when indulging in fast, fully laden test motoring. The small wheels seemed no serious handicap—except that they fitted potholes rather too well.

But supposing the driver wanted leisurely progress? Then, the engine would run down quite happily to a sedate 20 m.p.h. in top. The excellent and foolproof positive-stop hand-change gearbox retained its characteristics though the over-square motor might be labouring, for then the clutch freed 100%, as on all other occasions.

Not all three-wheelers, of course, have a reverse. The KR201 has four! By means of a reversing device in the facia ignition-cum-starter switch another contact-breaker is brought into use and the motor is automatically spun the " wrong " way. It then starts easily and runs, at full power, back-

wards, the rate of one's progress rearwards being limited only by the amount of throttle used and the gear in play! As rearward visibility is limited, this is strictly a parlour trick unless the canopy is lifted for full vision—a moment's job.

For serious use, the convertible was found to be quite capable of carrying two adults in comfort, and even a small child as well on the full-width back seat. A hingeing arrangement of the front seat aided the driver's entry and exit; leg room was adequate and all controls were easily reached. Handlebar steering is used and, being direct without reduction gear, is very precise; on first acquaintance the newcomer usually over-corrected at speed, but this tendency disappeared with experience.

A full specification is listed and includes twin headlamps and a good electric horn, an efficient silencer and an odourless situation for the fuel tank, its flow being controlled by an internal tap.

About the only criticisms of the convertible version of the *Kabinenroller* concerned detail points. A small window would be advantageous for signalling and ventilation, for the fold-back flexible side panels were not a complete answer. A better canopy-locking catch, accessible from inside and out, would be a boon, as also would more suitable dipped-beam characteristics for the headlamps, the standard layout cutting off rather too much light ahead.

However, these minor points aside, the Messerschmitt KR201 is a fast and nippy vehicle, easily parked and able to *more* than hold its own when energetically driven. Though its initial cost is not cheap, it earns bonus marks for its ability to transport two adults and a child with exceptional economy.

BRIEF SPECIFICATION

Engine: Sachs single-cylinder fan-cooled two-stroke; bore 65 mm. bv stroke 58 mm.; cast iron barrel; light-alloy cylinder head; c.r. 8 : 1; claimed b.h.p., 12; optional forward or reverse running.

Transmission: Four-speed unit-construction gearbox; positive-stop hand-change with neutral selection trigger; ratios, 4.2, 6.1, 9.1 and 17.7 : 1; final drive by enclosed chain.

Chassis: Tubular-steel welded construction.

Body: Soundproofed steel pressings; collapsible hood with folding stays; hinged canopy.

Wheels: Pressed-steel dummy-spoke type carrying 4-in. by 8-in. Michelin tyres; 4½-in.-dia. brakes, both hand and foot operated, on all wheels.

Lubrication: Petroil; test carried out with 24 : 1 ratio.

Electrical Equipment: 12-volt 135-watt Siba " Dynastart " starter-cum-generator; coil ignition with special contact-breaker and switching gear for running engine in reverse; twin 35/35-watt headlamps; wing pilot lamps and twin tail lamps; twin stoplamps; flashing indicators; electric horn; windscreen wiper; battery.

Suspension: Front wheels on stub axles pivoting on rubber bushes and controlled by spring units with hydraulic damping; pivoting engine-gearbox-rear-wheel unit working on gearbox mainshaft centreline; rubber rear suspension.

Tank: Welded steel of 3 gall. capacity; controlled by three-position tap with reserve position.

Dimensions: Wheelbase, 80 in.; track, 42 in.; ground clearance, 7 in. unladen; weight, 4½ cwt.

Finish: Bright-red glossy enamel; cream interior upholstery with black and white " lizard-skin " seat covers and trimmings in plastics material.

General Equipment: Full kit of tools; 80-m.p.h. speedometer; dummy aerial.

Price: £271 10s., plus £68 3s. 6d. P.T.= £339 13s. 6d. Extras: high-compression conversion, £11; sports silencer, £4.

Annual Tax: £5; quarterly, £1 7s. 6d.

Makers: FMR, Regensburg, Germany.

Concessionnaires: Cabin Scooters (Assemblies), Ltd., 80 George Street, London, W.1.

WITH a capacity for sustained cruising at over 50 m.p.h., a general fuel consumption of 65-70 m.p.g. and comfort comparable with that of a small car, the "two-and-a-half"-seater KR200 more than maintains the remarkable standards established by earlier examples of Messerschmitts' *Kabinenroller*.

It has a top speed on nodding acquaintance with 60 m.p.h.—a surprising figure for a three-wheeler propelled by a single-cylinder two-stroke of only 200 c.c. And a frisky engine, allied to a quick-changing four-speed gearbox, provides acceleration which is very far from sluggish. In this context, the low kerbside weight of 511 lb. is obviously important.

The KR200 carries its two adult occupants (plus a small child) in tandem under a side-hinged transparent canopy. The amount of room within the pressed-steel body appears to be limited and it is necessary to make use of the "raising-platform" seat device to slip readily into the driving position. But, once installed, both driver and passenger find that they have, paradoxically, more leg, arm and head-room than would be the case with the conventional small car, though vision is limited for a six-footer-plus by the upper framework of the canopy.

The addition of a child on the rear bench seat naturally limits the passenger's elbow room. Quite a lot of luggage may be carried.

The 191 c.c. Two-stroke
Messerschmitt KR200

A Three-wheeled Cabin Scooter Offering Economy, Comfort and Lively Performance

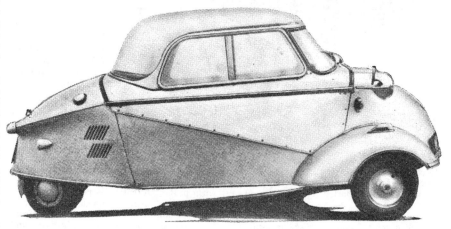

The "aircraft lines" of the KR200 are evident in side elevation. Despite its small overall height, the canopy provides ample head-room for tall adults.

(Below) Tandem cockpit accommodation: the controls are a mixture of car and motorcycle practice, with "handlebar" steering, pedal-operated clutch and positive-stop hand-change gear lever.

To prove the machine's comfort over a longish run, it was decided to go down A40, with sundry detours, leaving London about midnight, when the January air was below freezing, so as to test the standard heater.

With the "day's" mileage at Fishguard totalling 470, a number of points in the *Kabinenroller's* favour had been emphatically established. The most important related to the uncramped accommodation; the occupants were as relaxed as they would have been in the average 8 h.p. saloon car. And they were warm. To a driver far from fresh at the beginning of the run, the Messerschmitt proved an easy vehicle to drive. Interior acoustics were good; a normal conversation could be held without raising the voice.

Also clearly demonstrated was the good roadholding bestowed by the geometry and

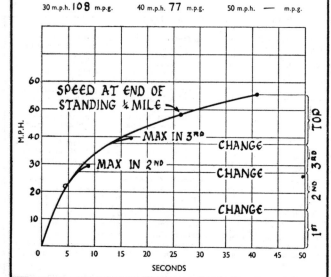

TESTER'S ROAD REPORT
MESSERSCHMITT KR200

Maximum Speeds in :—

				Time from Standing Start
Top Gear (Ratio 4·2 to 1)	56	m.p.h. = 5,200 r.p.m.	41	secs.
Third Gear (Ratio 6·2 to 1)	40	m.p.h. = 5,500 r.p.m.	17½	secs.
Second Gear (Ratio 9·1 to 1)	30	m.p.h. = 6,000 r.p.m.	8½	secs.

Speeds over measured Quarter Mile :—

Flying Start ___56___ m.p.h.   Standing Start ___34___ m.p.h.

Braking Figures On TARMACADAM **Surface, from 30 m.p.h.:—**

All Brakes 35 ft.

Fuel Consumption :—

30 m.p.h. 108 m.p.g.   40 m.p.h. 77 m.p.g.   50 m.p.h. — m.p.g.

SPEED AT END OF STANDING ¼ MILE

MAX IN 3RD — CHANGE
MAX IN 2ND — CHANGE
— CHANGE

M.P.H. (vertical axis: 10–60)
SECONDS (horizontal axis: 0–50)

suspension of the "two-in-front-and-one-behind" layout. The *Kabinenroller* showed no tendency to tip up, lift a wheel or break out of control if intelligently though quickly driven. When it did slide, it drifted with the back wheel first in the wet, and all three exactly together in the dry; correcting action was easily applied. Tyre-shrieking exercises could be performed without fear of the car gaining control of the driver. Steering at speed, however, required knack.

Through the Forest of Dean at 5 a.m., "raw" ice sheets crackled ominously under the wheels, but it was only necessary to ease the foot to maintain straight-ahead navigation. Past the Brecon Beacons, where packed snow and falling flakes tested the handling on tight bends, a combination of the right throttle and handlebar actions produced controlled slides; the tail did not wag the dog, though "road-happiness" at the back end could be felt.

The suspension stiffness necessary to minimize roll and retain good handling had not been overdone and an excellent measure of comfort was retained within the limitations imposed by the small wheels. Though the diminutive tyres were no drawback for encounters with normal bumps, they did fit pot-holes rather too well and the deeper depressions produced jarring. Handling was unimpaired by the addition of an eight-stone woman passenger and light luggage, and the springing was actually happier than in the "solo" condition.

Full throttle was held over the Oxford by-pass and the Cotswolds practically all the way to Andoversford, a distance of 35 miles. The little engine continued to produce sufficient power for the speedometer to indicate a consistent 58 m.p.h. on the level, with more in hand on downhill stretches. At 65 m.p.h. "indicated," the steering juddered, possibly due to unbalanced wheels.

With instrument correction, 58 m.p.h. "on the clock" represented a true maximum of 56 m.p.h. After allowing for a 3%-plus error of the mileage recorder, fuel on this fast part of the run was used at the rate of 57 m.p.g. More sedate usage brought consumption into the 75 m.p.g. category.

For 19 miles the *Kabinenroller* averaged 52 m.p.h. (!) and completed the whole of the fast journey to Pembrokeshire without stress. There was no need to clean the plug, adjust the brakes, take up the clutch or tinker with the push-pull mechanism of the positive-stop hand-change gear lever by the driver's right forearm. (The last-named needed firmness to ensure a clean change.)

Really powerful lights proved a blessing when approaching the occasional vehicle that was uncharitable in its response to a flashed appeal from WLF 916's twin headlamps. The foot-controlled dip-switch brought into play a pair of inoffensive beams that were bright and well focused.

The low driving position was a distinct handicap when the "opposition" did not dip. It was also a drawback in that, when the KR200 snuggled up behind a commercial vehicle, backwash from rain-sodden tyres was flung straight at the screen. Fortunately, the wiper was well up to its job and kept the glass free from dirt unless the roads were drying out; then a screen-washer would have been appreciated.

The brakes were fade-free, progressive and waterproof and could hardly have been bettered. The handbrake was not exactly an engineering masterpiece, though effective enough.

Pembrokeshire possesses some fine hills. The climb out of Fishguard Old Town on the Cardigan road is a test in itself, but

Even at speed in the gears, which normally meant changing up at 30 m.p.h. "indicated" in second and 40 in third, the box was completely silent. All changes were clash-free and quick. Clutch drag was non-existent.

The gearbox has no reverse. Reversing equipment takes the form of an alternative starter circuit which spins the engine backwards. It is brought into play by a facia switch; a green warning light identifies the reverse-selected condition. The test machine's electrics did not always willingly reverse the engine. On three occasions, they also refused to turn the engine in either direction after the machine had been standing out all night in some 10°F. of frost. No hand- or kick-starter is fitted. Generator output was excellent; pointing to a weak battery.

However, after a more normal night's rest in the garage, immediate starts were the rule. It was important to push the choke home as soon as firing commenced, indicating the possibility of a rich mixture for normal running. The plug, however, showed no sign of excess fuel.

Driver and passenger sit low in the pressed-steel body. The side-hinged canopy is weatherproof.

only second cog was needed, even when throttled back behind a lorry. From Dinas Cross, the by-road triple hairpin climb up towards the Yellow Mountain was a first-gear effort on the one-in-three corners. Nevertheless, a stop-and-restart test, with passenger, was no bother. It seems highly improbable that there is any metalled road which could defeat the KR200.

There were a few sources of minor annoyance. The 3-gal. tank capacity might be larger and the 70-mile reserve smaller. It is necessary to leave the side-hinged canopy open all the time the engine cover is up—a nuisance in wet weather. And one window of the test machine persisted in sliding open unless wedged. But these are small matters. All in all, the Messerschmitt *Kabinenroller* is a delightful vehicle, even for the dedicated motorcyclist brought up on two wheels.

---- BRIEF SPECIFICATION ----

Engine: 191 c.c. Sachs single-cylinder fan-cooled two-stroke; bore 65 mm. by stroke 58 mm.; cast iron barrel; light-alloy cylinder head; c.r. 8 : 1; claimed b.h.p., 12; optional forward or reverse running.

Transmission: Four-speed unit-construction gearbox; positive-stop hand-change with neutral selection trigger; ratios, 4.2, 6.1, 9.1 and 17.7 : 1; final drive by enclosed chain.

Chassis: Tubular-steel welded construction.

Body: Soundproofed steel pressings; transparent enclosure with safety-glass front-vision panel; hinging canopy.

Wheels: Pressed-steel disc type carrying 4-in. by 8-in. tyres; 4½-in.-dia. brakes, both hand and foot operated, on all wheels.

Lubrication. Petroil; test carried out with 24 : 1 ratio.

Electrical Equipment: 12-volt 135-watt Siba "Dynastart" starter-cum-generator; coil ignition with special contact-breaker and switching gear for running engine in reverse; twin 35/35-watt headlamps; wing pilot lamps and twin tail lamps; stop-lamp; flashing indicators; electric horn; windscreen wiper; battery.

Suspension: Front wheels on stub axles pivoting on rubber bushes and controlled by spring units with hydraulic damping; pivoting engine-gearbox-rear-wheel unit working on gearbox mainshaft centreline; rubber rear suspension.

Tank: Welded steel of 3 gal. capacity; controlled by three-position tap with reserve position.

Dimensions: Wheelbase, 80 in.; track, 42 in.; ground clearance, 7 in. unladen; certified kerbside weight, 511 lb.

Finish: Standard finish comprises one of eight basic colours; polychromatic versions available at £5 extra; test machine finished in silver-grey with red upholstery.

General Equipment: Kit of tools; 80-m.p.h. speedometer; heater; sliding windows; floor mat; bench-type back seat.

Price: £271 10s., plus £68 3s. 6d. P.T.= £339 13s. 6d.

Annual Tax: £5, quarterly, £1 7s. 6d.

Makers: FMR, Regensburg, Germany.

Concessionnaires: Cabin Scooters (Assemblies), Ltd., 80 George Street, London, W.1.

MISERLY MESSERSCHMITT

The makers say it's only a motor scooter with a cabin, some people say it's a car. Our road test shows that it's really both.

by PEDR DAVIS

IF you have not seen a Messerschmitt, you can gauge its size by knowing that it is seven inches less across the beam than the Austin A30; a Fiat 600 towers eight inches above it and a Customline is very nearly twice as long.

But don't let smiles of tolerance flicker into scorn. With one passenger aboard the Messerschmitt will come close to 60 m.p.h., its price tag is lower by far than anything else on the market — and its fuel consumption is exactly five times less than the Customline.

The Messerschmitt is the sort of vehicle people come to scoff at and remain to praise. During our extended test, we listened patiently to all the wags about the place. As we parked for lunch, a passer-by suggested that we chain it to the ground in case a housewife put it in her shopping basket. At a service station the bowser boy said "What do you feed it on—milk?" Later, a neighbour with a sports car looked the Messerschmitt over, asked its name and then chirped: "Say, how does it handle upside down in a cloud?"

A close examination, however, invariably provokes praise. The vehicle has been designed with immense ingenuity. The makers have packed into it almost every feature of a full sized car — including blinking lights and a heating system. Neither an ounce nor an inch have been wasted and consequently the running costs are unbelievably low.

For example, the registration tax is only £3 a year; the fuel economy runs around 90 miles per gallon; the routine maintenance costs are microscopic; chiefly because the wheels and steering are all suspended in rubber, so no periodic greasing (other than to the kingpins) are necessary.

The accommodation provides adequate room for two man-sized adults, plus extra space for a small child (or a large parcel) if required. In the closed version, complete protection from the weather is afforded and an added bonus is the fact that the Messerschmitt handles so well that you can play ducks and drakes in the traffic with impish ease.

Carrying two people, one behind the other is not a particularly friendly mode of transport, but that is how you ride in a Messerschmitt.

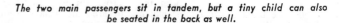

Apart from the handle bars, all the controls are of car type. Gear lever the right. For rearward motion, engine is reversed.

The two main passengers sit in tandem, but a tiny child can also be seated in the back as well.

Neither person is lacking in leg room nor head room and it comes as a major surprise to us to find it possible to sit as comfortably and as relaxed as we do in our office hack —which cost twice the price.

There is room for small parcels on a small shelf over the engine but this compartment is shut off once the bubble top is in place.

No one can complain of lack of room. The rear seat, being about eight inches wider than the front, provides parking space for an adult plus a small child, or two small children. Or you can ride solo — fold the seat up and carry up to two hundredweight of luggage.

Normally the rear passenger rides with one leg on either side of the driver. There are numerous small pockets for maps, gloves or cigarettes and even when two adults are carried, there is adequate room on the floor for sizeable packages. We took along a large camera case, an electronic flash unit, a Tapley meter, test equipment and two recording boards —and still felt there would have been room for more.

In order to make easier the entry and exit of the driver from his seat, an ingenious arrangement is resorted to. The front seat can be lifted upwards and backwards, over a pair of centre links, so that the driver's knee does not knock against the dash board. Once installed on the seat, the driver lowers himself forward and down into the normal driving posi-

tion. Very generous adjustment is provided so that the long and the short legged can place the seat just how they want it.

The bubble top can be lowered quietly and easily onto rubber buffers. It locks into place by means of a clamp and can also be secured from outside with a normal key.

The weather protection afforded by the bubble is first class. Visibility of course, is extraordinarily good, with very little distortion. Side windows slide to and fro, giving good ventilation without draughts.

One point, though, is that on a hot day the sunshine streaming through the bubble may be overwhelming and the passengers would feel happier equipped with hats or the optional extra canvas blind for the inside of the canopy.

Fortunately, there is ample headroom, even for a bowler.

Car type controls are used wherever possible and the starting procedure goes something like this: Turn on the fuel tap (next to the rear seat), then pull out the choke. Turn the ignition key. A combined starter-generator kicks the engine over and it purrs into life with the familiar pop-pop of a two-stroke. Press in the clutch and push the hand gear lever forward as far as it will go.

The gearbox is controlled with a "positive stop" system. That is, you change gear exactly as on a motor cycle. The lever is pushed as far for-

ward as possible to engage a gear. Then it returns to its normal position. Next gear is engaged by pushing forward again. You change down by pulling the lever back as far as it will go.

Neutral is selected by pressing a small trigger located on the lever itself.

Reverse is gained electrically, by reversing the direction of the engine and using the "forward gears" as reverse gears. To do this, it is necessary to stop the engine, press the key in and then turn. The engine then starts, spinning anti-clockwise.

The brake, clutch, throttle and dip switch are all operated by normal foot controls and the handbrake — a flimsy, rather noisy affair — is located on the left of the driver and too far forward for convenience.

The instruments consist of a 0-80 m.p.h. speedometer and four warning lights. They indicate the operation of the winking trafficators, high beam, ignition and the fact that the engine is spinning in reverse.

The steering "wheel" is a modern, nicely shaped handle bar. As it is linked to the front wheels without any gearing, steering is extremely direct. If takes only a shade of movement to deflect the car from its original path.

Now let's look at performance. First thing to note is that the tiny single cylinder engine is relatively quiet and commendably free from vibration. It purrs happily all through its speed range and provides a cruising range of about 40 m.p.h. with two adults aboard or 45 m.p.h. when driven solo.

The fuel consumption is literally what you care to make it. By driving with a feather foot on the throttle, a genuine 100 m.p.g. can be had. On the other hand, you can thrash the car up and down a test strip recording the performance figures and reduce the economy — as we did — to 65 m.p.g. Somewhere between the two lies the true mileage and we'd say that the average owner could bank on 75 m.p.g. for normal driving.

Being an air-cooled, two-stroke engine there are few maintenance worries. But it is important to carefully watch the tyre pressures. 15 lbs. for the front tyres and 30 lbs. for the single rear tyre is recommended — and unless these pressures are adhered to, some loss of road holding and stability will occur. It is interesting to note, by the way, that the special tyres needed for this car are now being manufactured in Australia —by Hardie.

Until the driver gets used to the direct, somewhat stiff steering, he is apt to consider the car unstable. But immediately full confidence in the machine is gained, you can really let your hair down. Since the maximum speed of the Messerschmitt is limited, there is very little need to slow down for corners. The car rolls round them with an almost breath-taking nonchalance and when rapid cornering does induce tail slide, it comes as a gentle reminder to the driver that he is taking things a little too close.

The steering is heavy, both at low and high speeds and certainly takes getting used to. The disappointing thing about the car is the turning circle of 30 feet. We feel this is an unnecessary handicap for an otherwise highly manoeuvrable vehicle.

A real crowd stopper, the Messerschmitt can fit into placs denied almost all other vehicles except motor cycles.

The suspension is firm but pleasing on smooth roads. When you get to pot holes it's best to ease up and take life leisurely. Otherwise the small wheels take a pounding and part of this at least is transmitted to the pasesngers. Even though the wheels are small, the ground clearance is ample at 6½ inches.

The car shows some tendency to roll at the front end on tight corners and this is probably one reason why the suspension is set on the firm side.

Frequent use of the gearbox is necessary for maximum performance —but it is worth noting that the clutch is light. Quick, easy gear shifts can be made in rapid succession. As on a motor cycle, it is best to change down as you approach a "Halt" sign.

so that an immediate start can be made in first.

Regular readers of "Wheels" road tests will note that we h a v e never yet tested a car that was entirely without fault. In the case of the Messerschmitt, the list of faults is comparatively small. First, we'd like to see an end to the present hand brake system and something more suitable installed. Whilst the plastic top gives excellent vision and removes any feeling of claustrophobia, we found it necessary to wear sun glasses for most of the time. The use of an anti-glare shield, placed directly over the driver's head would eliminate this criticism, without impairing the visibility. Quite a few big American cars have tinted plastic tops which work very well.

Another suggestion is that the bubble top should be hinged in such a fashion that it could be quickly detached from the car altogether. This would then provide the owner with an open car for use in the summer months. The way things are at present, the "convertible" model sells for some £10 less than the closed version — but has not the advantages of a fully enclosed cabin.

From a safety angle, the car scores highly. Its brakes are very powerful, the acceleration has a useful nip in it and the steering responds quickly to the helm. The only point here is that the car is so small that bus and truck drivers may be apt to overlook it.

Summing Up . . .

Make no mistake about the Messerschmitt. It is not a freak. It is a determined and praiseworthy attempt by a large German firm to build a vehicle that combines the best of two worlds — the scooter and the automobile. We consider the attempt has been justified. The vehicle has a very real place in the scheme of things. It is a pity though that it is classed as a car and therefore subject to a heavy sales tax slug.

Tiny, but very agile in traffic, the Messerschmitt stands little higher than the wheel of a double decker bus,

Specifications:

MAKE:
Messerschmitt three-wheeler cabin scooter; two passenger capacity. Test car from Messerschmitt (Australasia) Pty. Ltd., Sydney.

PRICE AND AVAILABILITY:
£548 (including tax). Availability: fairly good in most States at time of writing. (Convertible model £10 cheaper.)

ENGINE:
Single cylinder Sachs air cooled unit. Bore and stroke 65 mm. by 58 mm.; cubic capacity 191 c.c's; compression ratios 6.6 to 1; Bing piston valve carburettor; maximum power 10 b.h.p. at 5,250 r.p.m. Fuel tank capacity three gallons

TRANSMISSION:
Four disc plate clutch, motor cycle design; four forward gears with positive stop change mechanism Reverse selected by changing direction of engine rotation electrically. Gear shift by hand, via a Teledex cable. Final drive: chain to rear wheel. Overall ratios: Top 4.22; 3rd 6.06; 2nd 9.05; first 17.70.

CHASSIS AND BODY:
Tubular steel frame with all steel body. Side and bubble top form the only door, sliding side windows and wrap-around windscreen.

SUSPENSION:
Soft rubber torsion springs with shock absorbers on all three wheels.

BRAKES:
Cable operated internal expanding brakes. Foot brake and self locking hand brake acting on all three wheels.

ELECTRICAL EQUIPMENT:
12 volt system with coil ignition and 90 watt headlights. Electric windscreen wipers; flashing direction indicators; rear and stop light. Two 2 x 12 amp hour batteries connected in series.

STEERING:
Direct linkage, no steering box with handle bar control. Front wheels steered by divided track rods, rubber mounted.

OVERALL DIMENSIONS:
Wheelbase 6 ft. 8 in.; length 9 ft. 4 ins.; width 4 ft.; height 3 ft. 11 ins.; track (front) 3 ft. 6½ ins.; weight (unladen) 4¼ cwt.; weight full laden 7¾ cwt.; ground clearance 6½ ins.

Performance:

MAXIMUM SPEED:
Average of test runs, 56 m.p.h. (two passengers). Fastest one way run, 56 m.p.h.

MAXIMUM SPEED IN GEARS:
1st, 16 m.p.h.; 2nd, 29 m.p.h.; 3rd, 43 m.p.h.; Recommended shift points: 1st, 12 m.p.h.; 2nd, 22 m.p.h.; 3rd, 32 m.p.h. (normal driving).

ACCELERATION
Standing quarter mile: average of runs, 29.4 secs. Fastest one way run. 29.2 secs. (two passengers).

THROUGH GEARS:
0-20 m.p.h., 4.1 secs; 0-30 m.p.h., 9.8 secs.; 0-40 m.p.h, 18.8 secs; 0-50 m.p.h., 49.0 secs.

ROLLING ACCELERATION:
10-30 m.p.g. (2nd gear) 8.8 secs.; 20-40 m.p.g. (third gear) 14.0 secs.; 30-50 m.p.h. (top) 32.0 secs.

BRAKING:
Footbrake at 30 m.p.h. in neutral, 1.5 secs.; handbrake under same conditions 3.2 seconds; fade negligible.

SPEEDOMETER CALIBRATIONS:
10 m.p.h. (indicated) equals 10 m.p.h. (actual); 20 m.p.h.—18 m.p.h.; 30 m.p.h.—28 m.p.h.; 40 m.p.h.—38 m.p.h.; 50 m.p.h.—49 m.p.h.

TEST WEIGHT:
Driver, assistant, camera and test gear — 7¼ cwt.; weight distribution during test: front 3 cwt. 7 lbs.; rear 3 cwt. 35 lbs.

PETROL CONSUMPTION:
Normal driving 90 m.p.g.; driven hard including all performance tests, 65.2 m.p.g.

MESSERSCHMITT

The Messerschmitt KR 200 boasts a one-lung, two-cycle rear engine of 10 horsepower driving the single back wheel. It has tandem seating, a swingup top, rubber torsion springs and a cruising speed of over 50 miles per hour. Four forward gears are cable controlled; reverse is electrically selected. Final drive is a roller chain running in oil. The KR 200 sells at $998, and with a canvas top, at $1,014.

Only ten feet long, but four-wheeled and down-right frisky, the Messerschmitt TG 500 has a two-cylinder two-cycle rear engine of 20 horsepower. Independent coil-sprung back wheels are driven by articulated shafts. Front ones are rubber sprung. This minicar can get up to 60 miles per hour in 25 seconds, with a claimed top speed of 90. It is steered by tiller bar. East Coast price is $1,357.

By Bubble Car to Venice

SUGGESTION at a Sussex weekend party in April culminated in September, when seven men and four girls set out from London for Vienna in an assortment of miniature cars, accompanied by a professionally manned mobile film unit. The majority of us had saved up our fortnight's holiday, and it was in a mood of optimistic enthusiasm that we paraded at Admiralty Arch. The Scootacar, Nobel and Frisky had all covered less than a thousand miles; in contrast, the Messerschmitt had done over 27,000 miles, including a run to Rome and back, whilst the Heinkel had been in continuous personal use for a year and a half.

Mike Hawthorn gave the signal and the Bubbles rolled off into Trafalgar Square. As the cobbles and market stalls of the East End gave way to the roundabouts and dual carriageway of the

Ready to go aboard for the air-lift from Southend to Calais

We reached Basle at half-past six; the three-wheelers handled well on the tramlines, but the Swiss Highway Code baffled us for a time.

The precipitous descent into Lucerne for dinner on the next leg of our journey was considerably improved by grand opera from the Scootacar's Pye set.

It had been a long day and, as the result, none of us was up in time to see the sun rise over the Alps which we now faced.

One of our major difficulties was trying to go slow enough to allow the film unit, in a larger vehicle, to keep directly behind us, while at the same time keeping up sufficient speed to nego-

tiate the severe hairpin bends. Our air-cooled engines and short wheelbases proved a tremendous advantage, and by the time we reached Como at 9 p.m. we had covered 260 miles since midday.

Next day, thanks to the kindness of Signor Luppi, we were flagged off on four laps of the Monza track, and then set off for Milan. From then on most of our journey was done on autostrada, and this proved finally to our joy that these cars could withstand sustained cruising at near-maximum throttle without any ill effects. Verona was our next stop, and now we look forward to crossing the Dolomite range.

C. R. N. CARVALHO.

road to the coast, the cars found they were able to hold their own with the other traffic. Whereas we had expected that the average petrol consumption would be remarkably good, we now began to realize that we could also reckon on a surprisingly high average speed.

Embarkation at Southend Airport was scheduled for 6.30 p.m., but the light had already begun to fade when we taxied down the runway.

Formalities at Calais were in the best French tradition, and beyond the barrier we were greeted by Monsieur Denis, of B.P., Ltd., who led us off to a charming little hotel at Ardres, ten miles on our way.

On Sunday morning we were able to make considerable progress; in spite of a picnic which lasted rather longer than it should have, we were in Rheims soon after four o'clock. We dined at St. Dizier, and spent the night at Joinville, 20 miles farther on.

When we checked the cars next morning, we found that one tyre had been defeated by the *pavé* and that the silencer on the Messerschmitt had partially disintegrated. As this vehicle was in any event due for a replacement part at this stage, we sent it on ahead to the nearest agent at Basle, while we ourselves drove through some of the most attractive country we had yet come across.

Motor-minded Continentals take a good look at the little cars during a luncheon halt (above left)

A traffic check gives young cyclists an opportunity to see the cars in the convoy

The well-curved screen is in safety glass, but the sliding side windows and moulded roof are in plastic. The entire tail cowling can be removed after the spare wheel has been detached

LIKE a number of other miniature machines, the low priced, economical Messerschmitt three-wheeler has a keen following in Germany, its country of origin, and this has spread to some other European countries where there is much traffic congestion and difficulty in parking.

The F.M.R. Tg 500 four-wheeler from this factory, which uses the same basic layout, raised doubts as to whether space was being wasted by retaining tandem seating, since a full-width body with side-by-side seating could be adopted more readily in a design with four wheels. However, this could be judged only on the basis of the performance of the new model, because by retaining the narrow body width and low weight that goes with it, while increasing engine size to 490 c.c. (a vertical twin two-stroke Sachs engine is used), something approaching sports car standards of performance could be expected.

Rear suspension is independent, using a single wishbone and coil spring for each wheel; direct-acting hydraulically damped steering with no box, and with a handle-bar above an almost vertical column, is retained.

Control of the Tg 500 is something akin to that of a motor cycle combination, in that very small movements of the steering bar are required during normal driving, accompanied by a large amount of effort. The steering has a strange action, requiring a push-and-pull movement in a different direction from what might be expected in view of the angle of the control. Steering lock is disappointing for such a small vehicle, and at full deflection of the bar, the driver's hand comes up against the base of the windscreen. One important virtue, namely ease of manœuvring in confined spaces, is lost because of the large steering effort required. It makes reversing difficult, too, because both hands must be used to steer while looking to the rear.

Almost imperceptible movements are all that are needed to maintain a straight course on the road, but once the driver has become accustomed to the unusual action and

The unusual steering control on the Tg 500, with levers for operation of turn indicators, horn, and headlight dipping and flashing incorporated at its centre. Above the gear lever on the right is an ashtray

high gearing, he is able to place the car with a fair measure of accuracy, and having gained confidence can manœuvre with surprising speed. Like the three-wheeled version, this car is sensitive to side-winds, which easily throw it off its course, but with such direct steering, correction is very rapid. There is scarcely any self-centring action.

A notable feature of the car is the way it can be almost thrown round bends—the Tg 500 can be cornered very fast indeed with a complete absence of roll. There is a small amount of oversteer with one up, which increases when a passenger is carried and the fuel tank is full. When the car is taken beyond the limit of adhesion on a dry road, the rear wheels break away first, but correction can be made very quickly. With a full load, the car is less happy, feels rather tail-heavy, and develops a slight yawing motion at high speeds. This never gets out of hand, but is difficult to stop. In the wet, exuberance needs to be tempered with caution because rear-end break-away occurs much earlier. It is important that tyre pressures be raised by 3 lb sq in at the front and 5 lb sq in at the rear for high-speed driving, to give the best handling qualities.

Starting was always certain in cold weather, and, as in most two-stroke engines, the starter motor spins the engine for a few seconds before firing begins. The choke has to be held out during this time, but it can be released soon after the engine has fired. Then follows a short spell before the engine responds completely to the throttle and breaks into full song. The clutch was light and smooth, with a satisfactory amount of pedal travel, and no slip was experienced at any time during the test.

Less pleasing, however, was the gear change, which has no synchromesh, and at no time was there any great certainty about the movement of the lever, which is in two fore and aft planes as in a normal car. From standstill, first gear could never be selected without first slightly releasing the clutch, when the gear would engage with a noisy jolt. Quiet upward changes could be made only slowly, fast changes always being accompanied by some noise. Downward changes with this motor-cycle type box were made with the throttle held open, when they went through quite rapidly, but without that certainty of finding the gear which is so important, particularly if a change is being made before entering a corner. Reverse gear is engaged by a further movement of the lever across the gate towards the driver, followed by a forward movement, and it was easy to move the lever too far across the gate and find reverse when first was required.

The engine is a very free-revving unit, and quite impressive acceleration is possible if full use is made of the indirect gears. Acceleration figures are more affected in a small, light car such as this by the load carried, and some figures were repeated with the driver only in the car, to contrast with those recorded with the usual Road Test load. This improved the time for the standing quarter-mile from 22.8 to 22 sec, and the 0-50 m.p.h. figures from 16.7 to 14.5 sec.

Top gear ratio of 3.37 to 1 is very high, so that much motoring is done in third gear (4.75 to 1), which is also high for this type of car, the change to top being made only when on the open road. There is a large gap between second and third gear ratios.

In town, the normal practice is to remain in third, which gives sufficient acceleration unless speed falls very low, but when full use is made of this gear a marked engine vibration, which is felt throughout the car, begins at 50 m.p.h. on the speedometer (a true 43 m.p.h.) and continues until about 65 m.p.h. is indicated (57 m.p.h.). Beyond this the vibration smooths out a little but does not completely disappear. It is unfortunate that this should happen in the most useful part of the third gear range. The same vibration occurs at corresponding engine speeds in the other gears, but is then more easily avoided than in the much-used third gear ratio. The passenger is even more aware of this vibration than the driver, being closer to its source.

High top gear enables the car to cruise with some ease at 52-55 m.p.h., and 60 m.p.h. can be sustained, but with less comfort and confidence. A mean maximum speed of 65.5 m.p.h. was reached with two up, and with the driver only this was increased to 67.7 m.p.h., a highest reading of 70 m.p.h. being obtained. Speed falls off rather quickly when tackling the steeper main road hills in top gear, particularly with two passengers aboard, and third gear is usually engaged as soon as a hill is approached to maintain engine speed.

Engine noise is not really very great for an air-cooled two-stroke, probably because of its remote position in the tail, but conversation is rather difficult. Transmission noise can be heard on the overrun. Vibration, heavy steering and enclosure in a relatively small space combined to bring on tiredness on long journeys, and it was felt that this environment is more likely to be borne cheerfully by the young sports car enthusiast.

The engine could not be called flexible, pick-up from low speeds in third or top being poor. It was felt that there is some room for improvement in carburation to help these conditions. The engine did not four-stroke readily, and no oiling up of the plugs was experienced.

Hydraulic brakes have sufficient power for the performance, but tend to be rather fierce, and there is some front-end dip when they are applied hard. Quite moderate pedal pressures are called for, but there is a sense of instability when braking hard from high speeds, though the wheels do not lock easily. Awkward to reach and difficult to apply and release, the hand-brake nevertheless held the car on a steep hill, in spite of operation by cable on the front wheels only.

Riding comfort in the driving seat was good for a very small car, and improved when fully laden, but the rear seat occupant is less fortunate. Generally, springing is on the firm side, and there is a fair amount of short up and down movement, but the worst shocks are ironed out on poor surfaces and bumps do not affect the directional stability of the car, though severe ones are felt through the steering. On stone setts, there is some body rattle and drumming which is magnified by the shape of the plastic cockpit dome.

To enter the car, a normal door handle on the near-side is turned to release the canopy which, when closed, is retained by a substantial clamp around a frame tube. This handle can be locked by key from outside. The canopy, fairly heavy to lift, is supported in the fully open position

Side-lights and flashing turn indicators are in the front wings; rear-view mirrors are carried on the fairings of the high-mounted head lamps. The fog light and swivelling spotlight are not standard equipment

by a strap below the scuttle. The driving seat is mounted on four parallel-acting arms which allow it to be pulled back and up to make it more accessible for the driver. Once in the seat, he must then ease himself forward and lower the seat as gently as he can into the position for driving. This needs a little practice if the seat is not to fall too suddenly, and it is impossible to raise the seat before getting out. Stepping into either seat is made easier by the deep cut-away on the nearside of the body.

As the canopy is closed, a passenger on the rear seat must move his head forward slightly to prevent the rear edge of the canopy from striking him. An internal handle is then turned to lock the canopy in position. There is a surprising amount of headroom even for a tall driver, and a similar-sized occupant of the rear seat still has just sufficient headroom. The rear passenger, of course, has his legs on either side of the driving seat where there is ample space, but no step against which to brace his feet.

Apart from the steering bar and the hand-brake, all controls are of the normal car type. The brake and clutch pedals are spaced well apart, because the driver sits with his legs astride the steering column, and a feature that was much appreciated was the substantial foot rest on the left of the clutch pedal. In the absence of any kind of toe-board, this is valuable in helping the driver to brace himself. The gear lever is in the body side on the driver's right, and the only instrument is a speedometer (with total mileage recorder), which gave steady though very optimistic readings.

Out of sight behind the boss of the steering bar is the combined ignition and starter switch, operated by a key, and on the left of the facia—and difficult to reach because of the steering bar—are the choke and the lights switch. The hand brake, on the left at floor level, consists of a looped rod with a crude form of ratchet to retain it in the "on" position.

Controls for the flashing turn indicators, horn, head lamp dip switch and flasher, are operated by a pair of levers admirably placed one at each side of the steering bar centre.

The rear seat is just wide enough for a small child in addition to an adult. The smaller cushion can be removed to make room for a suitcase. Above the rear wing is the large filler for the 6½-gallon fuel tank behind the seat. The main part of the rear seat cushion can be raised and the smaller portion removed to leave the floor clear for extra luggage

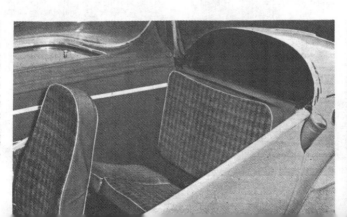

F.M.R. Tg 500 . . .

Extra tail lights to comply with British regulations are fitted on each side of the body outboard of the main lights. Flashing turn indicators are in the wings; the central reversing light is an extra fitting

body that he is inclined to forget that the wheel track is considerably wider. This applies also to the rear wheels, as it is easy to misjudge clearance when manœuvring close to other vehicles.

Misting-up inside is rather troublesome, as it is not possible to open the sliding side windows without some draught, and it was found necessary to have a cloth handy to clear the interior. Side windows are plastic; they become scratched rather readily and are too flexible to fit easily into the grooves at each end of their frame when closed, so that sealing is poor. A small opening of the front ones tends to produce wind roar. The plastic dome also is easily scratched when cleaned, and requires the use of a special polish periodically. The wiper swept a good area of the screen, but it was difficult to stop the blade in the parked position. In wet weather the car was spattered with mud from other vehicles more easily than with a normal car. A screen washer (an extra fitting) was found to be necessary in these conditions, but its jets reached only the lower edge of the screen at higher speeds.

The head lamps, set high in the body, have good penetration and a wide spread, and were adequate for the car's performance. In fog, the little car was particularly easy to drive because of its good visibility and the closeness of the driver to the screen and the left wheel. There is, however, a strong feeling of vulnerability, accentuated under these conditions, and with a vehicle of this size and low height other drivers, particularly those in large commercial vehicles, may find it difficult or even impossible to see the car if it is on their nearside. This has constantly to be borne in mind by the driver of a Tg 500, particularly when he is tempted to squeeze through small gaps in the traffic, a feat so easily accomplished with this car.

Access to the engine is rather restricted. There is a small hinged panel in the tail, locked by a key, through which the carburettor, plugs and twin coils can be reached. For more extensive work on the engine, transmission or rear suspension, the complete tail cowling with rear wings can be lifted clear after four bolts and the spare wheel have been removed. A further key unlocks a catch on the rear number plate, carried on a pair of hinged, plated arms, which can then be raised to reach the spare wheel (an optional extra) carried nearly vertically on the tail.

The fuel tank holds 6½ gallons, including three-quarters of a gallon reserve, which gives the car a range of about 270 miles between refills—a very useful capacity. Petroil mixture is used, in the ratio of half a pint of oil to 1½ gallons of premium petrol. Fuel consumption, which varied considerably with driving methods and conditions and the load car-

The left one is raised and lowered for head lamp dipping and main beam, and pressed for flashing the head lights, which can be done with the sidelights on or from the dipped beam position. The other control pivots for indicator operation, and is pressed to sound the horn. In practice, this was found an excellent arrangement, though with the controls above the steering arm, it was easy to catch a coat sleeve on one of the levers and move it inadvertently.

Although the squab of the driving seat is rather narrow, it is well curved and high enough to give support at shoulder level. It is quickly adjustable for rake in three positions. The cushion, too, though very firm, is properly shaped, rising towards the front to support the knees and along each side to locate the driver during cornering. Upholstery is in a woven non-slip p.v.c. material. There is a big range of fore and aft adjustment which allows drivers of any size to be correctly seated in relation to the pedals, and in the rearward position the front seat does not, of course, reduce the passenger's leg room. Elbow room, too, is sufficient, and a shut-in feeling is minimized because of the large transparent areas afforded by the curved safety glass screen, plastic side windows and domed roof which forms also the rear of the canopy.

Visibility is very good in all directions, with the reservation that it is not possible to see the front wings without peering through the side windows. This makes placing the car on the road more difficult than it should be, and the driver becomes so accustomed to the narrow width of the

Far left: With tail cowling off and air intake ducting removed, the position of the single Bing carburettor, sparking plugs and coils can be seen. Between engine and fuel tank is the 12-volt battery, and the large, flexible pipes above the engine are part of the heater system. Left: Locked by a key, this small panel contains the carburettor air intake duct and gives access to the upper part of the engine. In the centre is the air filter

ried, was between 41 and 58 m.p.g., the lower figure including hard, open road driving and congested traffic.

The rear seat is wide enough for a child to sit beside an adult for short journeys, and the cushion, which is in two parts, can be hinged upwards to enable the whole of the rear of the cockpit to be used for luggage. There is, however, a shallow boot under the tail panel, reached from the cockpit when the canopy is open.

A simple heater—an optional extra—which picks up warmed air from the ducts around the cylinders and leads it through flexible pipes to points near the driver's right foot and beneath the rear seat, gives insufficient warmth in cold weather, but a branch pipe to the windscreen helps to reduce misting; the control is on the near bulkhead beside the seat squab. There is a comprehensive set of good quality tools; a spare driving belt for the engine cooling impeller and a normal screw jack are included in the kit. Sole U.K. concessionaires are Cabin Scooters (Assemblies), Ltd., 80, George Street, London, W.1.

It would seem that the Tg 500 must have a limited appeal in this country at a price swollen by import duty and purchase tax to over £650. Performance, particularly acceleration, is good for a 500 c.c. car, but few sports car enthusiasts would approve of the steering or gear change. Although tandem seating is acceptable in a low-price economy runabout, a less freakish and more spacious layout is required for the serious motoring that is expected from a four-wheeler.

F.M.R. Tg 500

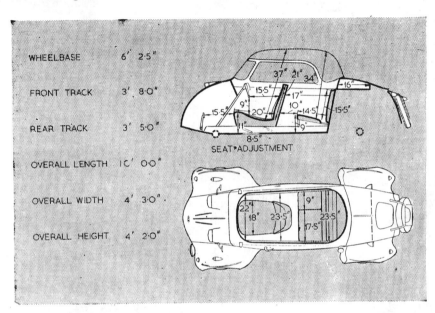

WHEELBASE	6' 2.5"
FRONT TRACK	3' 8.0"
REAR TRACK	3' 5.0"
OVERALL LENGTH	10' 0.0"
OVERALL WIDTH	4' 3.0"
OVERALL HEIGHT	4' 2.0"

Scale ⅛in to 1ft. Driving seat in central position. Cushions uncompressed.

PERFORMANCE

ACCELERATION:
Speed range, Gear Ratios and Time in sec.

M.P.H.	3.37 to 1	4.75 to 1	8.22 to 1	15.2 to 1
10—30..	—	—	5.9	—
20—40..	—	11.5	—	7.1
30—50..	23.4	12.0	—	—
40—60..	—	16.6	—	—

From rest through gears to:

M.P.H.	sec.
30	5.9
40	10.0
50	16.7
60	27.8

Standing start quarter mile 22.8 sec.

MAXIMUM SPEEDS ON GEARS:

Gear		M.P.H.	K.P.H.
Top	(mean)	65.5	105.4
	(best)	68.0	109.4
3rd		62	99.8
2nd		42	67.6
1st		24	38.6

TRACTIVE EFFORT:

	Pull (lb per ton)	Equivalent Gradient
Top	97	1 in 23.1
Third	192	1 in 11.6
Second	362	1 in 6.1

SPEEDOMETER CORRECTION: M.P.H.

Car speedometer:	10	20	30	40	50	60	70
True speed:	9	17	26	34	43	52	61

BRAKES: (at 30 m.p.h. in neutral):

Pedal Load in lb	Retardation	Equivalent stopping distance in ft.
25	0.25g	121
50	0.59g	51
65	0.71g	42.6

FUEL CONSUMPTION:
M.P.G. at steady speeds

M.P.H.	Top
30	71.4
40	66.6
50	57.1
60	37.1

Overall fuel consumption for 824 miles, 43.8 m.p.g. (6.5 litres per 100 km.).
Approximate normal range 41—58 m.p.g. (6.9—4.9 litres per 100 km.).
Fuel: Premium grade.

TEST CONDITIONS: Weather:
overcast, wet tarmacadam road. Wind speed 0-10 m.p.h. Air temperature, 46 deg. F. Acceleration figures are the means of several runs in opposite directions.
Tractive effort obtained by Tapley meter.

─── DATA ───

PRICE (basic), with saloon body, £435.
British purchase tax, £216 5s. 8d.
Total (in Great Britain), £651 5s 8d.
Extras (including purchase tax): Radio £22 10s. Heater £5 15s. Conversion to cabriolet roof £6 10s. Spare wheel £5 5s.
ENGINE: Capacity, 490 c.c. (7.27 cu in). Number of cylinders, 2 in line (transverse). Bore and stroke, 67 × 70 mm (2.64 × 2.76in). Valve gear, ported two-stroke. Compression ratio, 6.5 to 1.
B.H.P. 20 (nett), 24 (gross) at 5,000 r.p.m. (B.H.P. nett per ton laden 37.9).
Torque, 24.6lb ft at 4,000 r.p.m.
M.P.H. per 1,000 r.p.m. in top gear, 18.
WEIGHT: (With 5 gals fuel), 7.56 cwt (847lb).
Weight distribution (per cent); F, 29; R, 71.
Laden as tested, 10.56 cwt. (1,183lb).
Lb per c.c. (laden), 2.42.
BRAKES: Type, V. Knott hydraulic.
Drum dimensions; F, 7.1in diameter; 1.2in wide. R, 7.1in diameter; 1.2in wide.
Lining area: F, 31 sq in; R, 31 sq in (117.5 sq in per ton laden).
TYRES: 4.40—10in. Metzeler.
Pressures (lb sq in); F, 15; R, 25 (normal). F, 18; R, 30 (fast driving).
TANK CAPACITY: 6.5 Imperial gallons (includes 0.75 gallon reserve).
STEERING: Turning circle:
Between kerbs, L, 30ft 7in; R. 28ft. 0in.
Between walls, L, 31ft. 4in; R, 28ft. 10in.
Turns of steering bar, lock to lock, 0.3.
DIMENSIONS: Wheelbase, 6ft 2.5in.
Track: F, 3ft 8in; R, 3ft 5in.
Length (overall), 10ft.
Width, 4ft 3in.
Height: 4ft 2in.
Ground clearance, 6in.
Frontal area, 10.7sq ft.
ELECTRICAL SYSTEM: 12-volt; 24 ampère-hour battery.
Headlights, double dip: 35—35 watt bulbs.
SUSPENSION: Front, independent, single wishbones and rubber in torsion. Telescopic dampers. Rear: single wishbones and coil springs. Telescopic dampers.

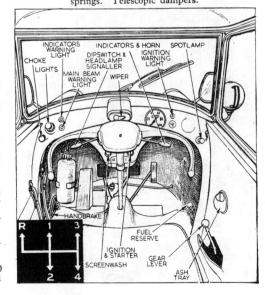

F·M·R

Tg 500

The phenominal SPORTSCAR

The latest,
the best
The greatest little car!

What do you expect of a sportscar?

a   Easy access to engine

b   Phenominal acceleration

c   All-round visibility

d   Stability when cornering

e   Excellent climbing power

f   Manoeuverability in traffic

g   Instantly responsive steering

h   Reliable brakes

i   Well-balanced weight distribution

k   Economical fuel consumption

l   Sturdy body

m   Superior roadholding

n   Comfortable seating

o   Superb headlights

The Tg 500 fulfills your wishes!

a Removable engine cowling · all engine parts easily accessible · simple lay-out for servicing · small vent obviates removal of cowling for access to plugs, carburettor, ignition coils, regulator, petrol pump and filter.

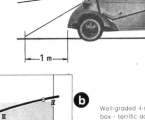

c Panoramic windscreen giving unexcelled all-round vision.

— 1 m —

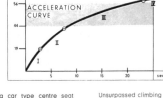

MPH

56

44

19

ACCELERATION CURVE

I II III IV

5 10 15 20   sec.

b Well-graded 4-speed gear box · terrific acceleration, reaching 65 mph in 3rd gear.

d Racing car type centre seat exceptionally low centre of gravity making overturning impossible · Minimum body sway in cornering.

e Unsurpassed climbing ability, even with full load on mountain gradients.

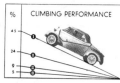

% CLIMBING PERFORMANCE

4.5
24
9
5

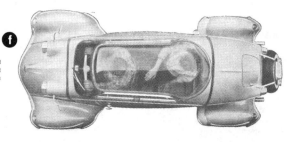

f Sleek body shape, yet ample elbow room · utmost safety in narrow streets and dense traffic.

Touring

High cruising averages ·
comfortable seating

In Traffic

Nippy · manoeuverable · easy to park

In the Country

Robust · dependable · powerful

Also available as Convertible

Technical data for Tg 500

Engine

Type FMR 500 L; Twin cylinder, Two stroke, 490 cc; Bore 67 Mm, Stroke 70 Mm, Compression ratio 1:6.5: Performance 20 BHP at 5,000 r. p. m. cooling by two fans separate for each cylinder; 3 Stage Bing Carburettor; special moisture proof Air Filter; fuel supply by Fuel Pump; 18 Mm sparking plug, Heating capacity 225; Bosch Dyna-Starter, 12 V / 160 W.

Tank Capacity

6½ Gallons of which ¾ gallon is reserve - Reserve tank control on dashboard.

Transmission

Gearbox in Engine block, 4 forward and 1 reverse gear - ENGINE-Transmission ratio 1.82; 2 Plate dry clutch; final drive to rear wheels through spur wheel and differential to articulate drive shafts. Gear reductions 1st 15.2; 2nd 8.22; 3rd 4.75; 4th 3.37; Reverse 15.82.

Frame

Torsion - Proof tubular steel frame with completely closed floor.

Body

All steel body with weather resistant furnace set coat of paint.

Suspension

Independent suspension with progressive springing, self-acting stabilising control on rear wheels; hydraulic shock absorbers; torsion rubber sprung front wheels, coil sprung rear wheels.

Steering

Direct steering by wear-proof divided track rods with synthetic bearings; hydraulic steering damper; modern steering bar with finger tip levers for dipper control, overtaking flasher and trafficators.

Brakes

Hydraulic footbrake acting on all four wheels. Mechanical hand brake for front wheel, 180 mm. diam.

Tyres

4.40x10 High velocity sports tyres. Spare wheel locked to engine cowling. Wheel base 74⅗ inches, track width front 44 inches, rear 41 inches.

Interior

Entry by lifting hinged dome structure; Tandem seating, foam rubber upholstery, Front seat fully adjustable, full width removable back seat, accommodating one adult and one child.

Electrical Equipment

12 V / 24 Aph Battery, High set head-lights, 5 inch diameter beam-dipper, overtaking flasher, large number plate lighting rear lights with reflectors and stop light. Trafficators on front and rear wings with dash-board signal, windscreen wiper, panel lights.

Dimensions

Length 120 inches, width 51 inches, height 50 inches, ground clearance 6 inches, unloaden weigth 770 lbs, turning circle 31 feet.

Performance

Acceleration: 0—44 m.p.h. 7.2 seconds., 0—50 m.p.h. 14.5 seconds., 0—62 m.p.h. 25 seconds: Top speed approximately 90 m.p.h., Climbing performance 1st gear 45%, 2nd gear 24%, 3rd gear 9%, 4th gear 5%; Fuel consumption 52 m.p.g'

FMR FAHRZEUG- UND MASCHINENBAU GMBH REGENSBURG

g Featherlike steering · highly manoeuverable · astonishing control of vehicle · comfortable steering bar · fingertip levers for dipper, overtaking flasher, trafficators and horn on steering bar · hydraulic steering damper.

h Large-sized hydraulic brakes on all wheels with smooth pedal action.

Passenger sits in centre of gravity of vehicle · weight distribution in either direction independent of passenger · ideally balanced weight distribution prevents drift and sway.

k Aerodynamic styling and light-weight construction of the body results in minimum fuel consumption.

Selfsupporting shell with torsion-proof tubular steel frame.

m Independent rear-wheel suspension with self-acting telescopic stabilisers · coilsprings over hydraulic shock-absorbers · wheels individually balanced.

n Racing car type bucket seat with fully adjustable backrest.

o Powerful headlights · modern assymetrical dipperlights · added safety through wide forward thrust of beam and good illumination of kerb.

ROAD TEST

65 MPH AND 80 MPG—THAT'S THE

MESSERSCHMITT

MOTORCYCLE MECHANICS ROAD TEST No. 54

Vehicle Messerschmitt KR200 Sports Saloon **Price new** £348-16-9

Engine Sachs 191cc Fan cooled, two stroke 9.7 bhp @ 5000 rpm

Gearbox 4 Speeds forward and reverse, hand lever positive stop

Final drive Chain in oil bath

GENERAL INFORMATION

Weight	506 lbs
Saddle height	—
Turning circle	27 ft.
Is toolbox lockable	Yes
Is steering lockable	Yes
Fuel tank capacity	3 gals
Reserve capacity	1/2 gal
Oil tank capacity	—
Gearbox capacity	1 pt approx
Fuel specified	Commercial petrol
Overall consumption	81 mpg
Braking from 30 mph	39 ft
Acceleration 0-50 mph	27 secs

SPEEDS IN GEARS

(bar chart, mph vs GEAR 1–6)

EQUIPMENT SUPPLIED

STANDARD FITTINGS Winking indicators, Ignition warning light, mirror

OPTIONAL EXTRAS Spare wheel, luggage rack, bucket seat, visors, clock, wheel discs, lockable tank cap

SPARES PRICES

Engine gasket set	with oil seals 13s 6d
Set valves & guides	—
Piston with rings	81s 3d
Set of clutch plates	31s 4d
Silencer	96s 6d
Pr Exchange brake shoes	15s 2d

RATING CHART
(points out of 10)

Control positioning and adjustment	7
Instruments and equipment	5
Fuel reserve and tap operation	7
Ease of starting	7
Engine smoothness	7
Quietness of engine & transmission	6
Gearbox and clutch operation	6
Road holding	8
Braking efficiency	5
Comfort and ease of handling	8
Lighting efficiency	8
hand brake stand operation	5
Tool kit	7
Overall finish	8
General performance & reliability	9
TOTAL (maximum points 150)	103

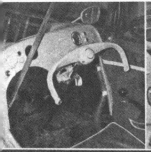
Perspex dome of the roof back and is held by strap. aeroplane-type steering

Driver's seat is small but still comfortable. Two people can be squeezed into the rear seat

If only one passenger is in the car, part of the seat lifts up, leaving room for a case

In order to make things easier when getting in the car, front seat lifts about six inches

ools are the same as in mal car. Dipswitch is to eft of the clutch pedal

The blower-cooled 191 c.c. two-stroke Sachs engine is at the rear under a large alloy cover

The Messerschmitt has quite a reasonable toolkit which even includes touch-up paint

Luggage-carrier is a useful extra fitted to rear of car. Takes heavy suitcase with ease

UNTIL I tried our road-test Messerschmitt KR 200 I would never have believed it possible to form three entirely conflicting opinions of one machine in a single afternoon. However, only two or three hours after picking up the 200 c.c. three-wheeler from Two-Strokes Ltd., of Stanmore, that was exactly what I had done.

First impression was one of scorn. This bubble-car has a beetle-like outline that is not exactly pretty, and to climb into it you have to lift up the roof ! After I had entered the car, however, this impression soon vanished. I had been told that the Messerschmitt Kabinen Roller (it means Cabin Scooter) was based on the cockpit of the wartime ME109 and on sitting inside it I found this easy to believe. There was no steering wheel, but instead I was faced with an aeroplane-type joystick.

Looking through the glass dome roof at the smooth contours of the body I realised just how well-streamlined it was. With the passenger sitting behind the driver, frontal area had been reduced to a minimum, thus giving better penetration and its attendant advantages of more speed and better fuel consumption figures. The car-type controls were well-placed but a little close together—only to be expected with a vehicle as narrow as the Messerschmitt. The driver's seat was small but comfortable and hinged up so that getting aboard was not quite as difficult as it first appeared.

Once I was installed in the " flying " position, I took off for the open road— well the Edgware Road, anyway—and that was where I got my second impres-

sion . . . one of sheer terror ! The steering column is connected direct to the front axle and as a result the slightest movement of the handlebars meant a great sweeping swerve across the road. After frightening myself and several oncoming car-drivers almost to death, I at last mastered the technique and found that there was no need to grip the handlebars. They could just be held lightly and the car would practically steer itself. Bumpy roads, however, presented a problem. The Messerschmitt is so light that any big bumps really sent it skywards and I found that it was always best to be ready to deal with this rather than be caught unawares and maybe end up in the ditch.

It was on a rush-hour trip up Western Avenue at the start of a 70-mile journey to Banbury, in Oxfordshire, that I got my third and lasting impression of the little car. Not one of scorn or of terror—but of pure joy. All the way around the Western Avenue roundabouts I was dicing merrily with all sizes of cars ranging from Mini-Minors to 3 litre Rovers. On the straights, of course, I was left languishing in the distance but on the icy surfaces of the many corners the road-holding of the Messerschmitt was at least equal to that of anything else in the vicinity.

Just after I had taken the bubble-car to Banbury, the January blizzards started in earnest and it was during this period that I became completely converted to the Messerschmitt faith. In all the time that I had the car on test, thick snow lay on the ground and its handleability under these conditions was absolutely phenomenal. During that time, in fact, the number of vehicles that passed me on the open road could literally be counted on

my fingers.

On one trip through Arctic conditions, almost to High Wycombe and then back to Banbury, I passed everything in sight and was only repassed by one lorry . . . on a comparatively dry stretch of road. Down a long hill I pushed the speed up to 60 m.p.h. on a surface of ice covered with a thin dusting of snow. The rear of the car was swinging mightily from side to side but the front was still under perfect control. Judging by the expressions of several bystanders, it may have looked rather hairy but I never once felt any cause for alarm.

I got so used to the handling of the car, that I was taking snow-covered corners like a dervish and going round in go-kart fashion. In fact, I enjoyed watching pedestrians' amazement as they watched the Messerschmitt cornering almost like a racing car, while other vehicles were being driven at crawling pace.

While handling was, without doubt, the car's strong point, speed and economy were also well to the top of the list. Top speed was 67 m.p.h.—equal to a good many motorcycles of similar capacity and average fuel consumption was a shade over 80 m.p.g.

Braking, however, was not so good. It took 39 feet to stop from 30 m.p.h. and on long journeys the brakes faded and became progressively worse. A point to bear in mind here, however, was that our test model was second-hand and those of you who buy new Messerschmitts should be able to expect better brakes than this.

Though usually a two-seater, the car could carry three, provided the two in the rear were not too big. Seating was quite

Continued on page 104

MESSERSCHMiTT KR 200

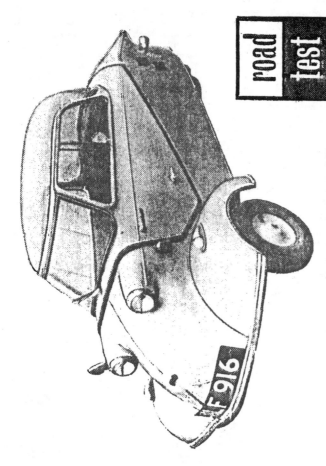

She's a joy to handle in traffic

THE general design features of the Messerschmitt three wheeler are well known even to the relatively disinterested observer. This is due to the "different" appearance, although this is not to say the Messerschmitt is an oddity. The overall design is good and the transparent cabin section is well balanced in relation to the body size.

The Messerschmitt is light. On a flat road you find that it can be pushed along as easily as a baby-carriage. With a bit of an effort the front can be lifted up by one person, with a little more effort the rear wheel can be raised. With such low rolling resistance, and low air resistance owing to the small frontal area and excellent aerodynamic form of the attractively shaped body, the power output of the two-stroke 191 c.c. engine is used to best advantage. An indicated top speed of 70 m.p.h. was, in fact, obtained solo, with the KR 200 submitted for test.

At this speed the steering feels rigid.

Road shocks are ironed out by the suspension to the extent that they can be heard rather than felt. The noise of the tyres on the road surface, the hiss of the slipstream and the pleasant note of the motor at full song make such speeds an exhilarating experience.

The machine feels as much in contact with the road surface as it does at 40 m.p.h. There is little sway and no feeling of being "airborne". Any doubts one might have concerning the stability of the vehicle at speed because of the lightness of its construction appear groundless.

In testing any vehicle one must, for part of the time, assume the approach of somebody who is completely new to the particular machine on trial.

In applying this test to the Messerschmitt one discovers that the newcomer may have doubts as to its directional stability. To understand why we must take a look at the system of steering used.

This means, firstly, that the slightest movement of the steering bar gives a corresponding movement to the front wheels. Secondly, that the steering is not geared down and therefore feels heavy.

The combined effect is likely to make a new driver hold the steering bar tightly and make definite movements to correct any alteration that may momentarily occur to the vehicle's course as the result of, say, a sudden change of road camber.

The result is that he overcorrects and a certain amount of snaking commences. If he doesn't realise what he is doing, his first impressions of the Messerschmitt's directional stability may not be good.

A relaxed grip of the steering bar, gentle but quickly applied pressures, not movements, will take care of even severe changes of camber and surface, leaving actual movements of the steering bar for the tighter corners. As speed is increased the steering becomes even harder and steadier, and over-correction is less likely. These steering characteristics can be quickly mastered with a little practice.

Handling such a small vehicle in town is a joy. The rapid changes of direction possible, the size of the vehicle, the sprightly acceleration, and the ease with which gear changes can be made, allow you to slip through the traffic. No matter

The engine can be exposed for maintenance purposes by opening the hood and swinging the rear casing upwards.

Steering is direct. Power from the wrists, applied to the attractively styled steering bar, is transmitted by a steering column to a short lever welded at 90 deg. to the column at floor level. From the free end of the lever pivot the two arms of a divided track rod. Each arm runs outwards to left and right to turn the front wheels as the steering bar is turned.

DATA PANEL

MESSERSCHMiTT KR 200 DE LUXE

Engine. Sachs 191 c.c. blower-cooled single-cylinder two-stroke; bore 65 m.m.; stroke 58 m.m., compression ratio 6.6:1, claimed b.h.p. 9.7, 5,250 r.p.m.

Carburettor. Bing; air-intake silencer and filter.

Transmission. Four-speed unit construction gearbox; positive-stop hand-change with neutral selection trigger; 4-disc clutch; final drive, enclosed chain.

Chassis. Tubular-steel, welded construction.

Body. Steel pressings. Hinged hood; Perspex dome and sliding windows; curved windscreen, safety glass.

Wheels. Steel rims, ornamental discs. Tyres 4.00 × 8.

Lubrication. Petroil 24:1 mixture.

Electrical equipment. 12-volt lighting, Bosch Dynastart, 135 watts; coil ignition, special contact-breaker and switching gear for running engine in reverse; 35/35-watt headlamps, wing pilot lamps and twin tail lamps; twin stop lamps; flashing indicators; electric horn; windscreen wiper; battery.

Suspension. Independent; soft rubber torsion springs with hydraulic shock absorbers on all three wheels.

Tank. Welded steel. 3-gallon capacity. half-gallon reserve.

General equipment. Full kit of tools, m.p.h. speedometer.

Price. £344 12s. 3d. inclusive tax.

Extras. Special metallic finishes. Spare wheel. Roll-back canvas hood instead of Perspex dome. Sports silencer.

Makers. FMR. Regensburg. Germany.

Concessionaires. Cabin Scooters (Assemblies) Ltd., 11 South Wharf Road, London, W.2.

Dimensions: Overall length (A)—9ft. 4in. Overall height (B)—3ft. 11in. Overall width—4ft.

The cockpit (the term seems to come quite naturally when talking about the Messerschmitt KR 200) is entered by raising the transparent plastic hood.

wiper keeps the screen clear in wet weather. The interior is red leatherette with white plastic trimming. Black rubber covers the floor.

The KR 200 is easily cleaned. Below the vehicle is a flat floor, ribbed at intervals for strength, and a model of good weather protection.

With the hood open to get in or out, the whole of the rear casing can be swung upwards to reveal the engine and rear suspension for maintenance and inspection. Rear suspension is trailing arm with rubber-in-torsion for springing. Hydraulic shock absorbers are fitted.

full beam leaves nothing to be desired. The "dip" position, however, cuts the beam off too close to the vehicle to be comfortable. Though the level of illumination within the lit area is excellent, it would be nice to be able to see a little farther ahead, particularly against the lights of an approaching vehicle.

Reversing is possible by stopping the 191 c.c. Sachs engine and starting it again in a reverse direction. This is accomplished by means of the Bosch Dynastart and ignition system. The ignition key is inserted in the lock and turned to the right to switch on. Turning the key farther to the right against a return spring operates the starter in the forwards direction.

If the key is inserted in the lock and then pushed and held in as it is turned to the right, a second and reverse ignition system is brought into operation, and the green indicator light comes on as well as the normal red one.

If the key is now turned farther to the right, as before, against a return spring, the engine is started in the reverse direction, and both indicator lights extinguish if the engine runs above much more than tick-over speed. It is of course necessary to switch off and wait until the engine has stopped before operating the starter in the other direction.

Engine and Gearbox

The Sachs engine and gearbox are in unit construction and they do not move with the trailing arm. The final transmission is by fully-enclosed chain in an oil bath, and transmission from the gearbox to the driving sprocket is by Cardan shaft. The arm pivots about the line of the shaft so that there is no change in chain tension as the rear wheel deflects under varying loads.

Speeds in the gears were: 1st gear, from almost tick-over to 20 m.p.h.; 2nd, from 9 m.p.h. to 34 m.p.h.; 3rd, from 12 to 47 m.p.h.; 4th, from 28 to 70 m.p.h. on a flat road under windless conditions. The lower figures were the minimum obtainable without transmission snatch, but the engine is not really happy under 30 m.p.h. in top gear.

Cruising between 30 m.p.h. and 50 m.p.h. is a pleasurable experience, the engine making light work of such speeds. One-in-three hills can be climbed at 30 m.p.h. in third gear, though with a full payload this figure drops a few m.p.h. so that it is kinder to the engine to drop to a lower gear. Less steep hills are climbed with ease, with plenty of power in reserve.

The front wheels are also independently suspended and damped with hydraulic shock absorbers. Deliberate drifts on very sharp corners were readily controlled, and any tendency to breakaway occurs at the rear wheel—a safe state of affairs. On main roads the small wheels seem no handicap at all. Sedate cruising at a steady 40 m.p.h. on country roads with the Messerschmitt KR 200 proved most enjoyable, particularly when a petrol consumption figure of 87.2 miles per gallon was recorded.

Petrol Tap and Heater

At the back to the offside of the rear seat is the petrol tap and the heater control knob. The capacity of the tank is 3 gallons, and a half a gallon of this is held in reserve and brought into use by turning the petrol tap to the "reserve" position. The heater knob is pulled out to bring a hot air blower system into operation. Engine heat is used to warm the air which is taken from the blower fan casing on the engine, and conveyed to the cabin through a flexible hose. The heat is appreciable and the interior is warm in cold weather.

Leg room for the rear passenger is obtained by pushing the legs past the driver's seat. Part of the rear seat is divided to provide a seat for a small child, or it can be folded upwards to provide luggage room. Both seats fold upwards to provide even more luggage room when the machine is used solo.

Sliding Perspex windows take care of the ventilation and a good windscreen

how the accelerator is used, the exhaust note is always pleasant, and commendably quiet if the accelerator is used with care. Parking is easy, though it may be a surprise to find that the turning circle at full lock is approximately 27ft. 2in., measured from the inside wheelhub.

Four gears are provided in the range, and all could be used in reverse. The gear change lever is at the right hand, and consists of a simple lever which is moved backwards for changing down, forwards for changing up.

Each movement has a positive stop, and each full pull or push selects the next lowest or highest gear respectively. Though there is a certain feeling of coarseness in the movement, gear selection is very positive, provided that a full and deliberate movement is made. However badly the engine and clutch is mishandled there is no complaint from the gearbox.

Neutral selection is made by means of a small trigger on the gear lever. With the clutch depressed this trigger is operated to shift the gears into the next adjacent neutral position between the gears in use. From fourth gear, for example, neutral is obtained between the third and fourth gears.

By merely squeezing the trigger, neutral can be selected at any speed, from any gear, with the clutch depressed. Before starting again, several backward pulls at the gear lever ensure that bottom is engaged, since if neutral happened to fall between second and third, the next backward pull of the gear lever would select second gear, and one more pull would be needed to obtain bottom gear.

The remainder of the controls are car-type. On the floor to the extreme left is the foot operated dip-switch, and nearer the centre, by the steering column, the clutch pedal. On the right of the column is the brake pedal and on the extreme right the accelerator. Above foot level at the left is the hand-brake. This is very effective and it applies the brakes on all three wheels. The foot brake also operates on all three wheels and the test machine gave braking efficiencies of 60 per cent, a very good figure for this class of vehicle.

On the instrument dash-board car-type controls are also fitted. At the extreme left there is the choke control. This is an easy-start device which is released as soon as the engine starts. Next to it is the light switch, and high-beam warning light.

The lighting system is 12 volts, and the light produced by the headlamps at

Messerschmitt

motor-car comfort at motor-cycle costs!

There's motor-car comfort because the close-fitting roof makes you feel delightfully snug in your smart *Messerschmitt*, sealing you off completely from rain and cold weather; motor-cycle costs because its robust 200 c.c. two-stroke engine produces a fuel consumption of 87 m.p.g.; the annual tax is only £5; and the insurance as low as £4.10.0. In fact, it's cheaper than travelling by bus or train.

plus superb performance

1 *Speed.* The *Messerschmitt's* unique aerodynamic 'tandem' design reduces wind resistance, thereby not only lowering running costs but also raising its performance tremendously. It is the fastest of all small cars, with a top speed of 67 m.p.h. Nippy in traffic in heavily populated areas and also in rural lanes.

2 *Stability and Roadholding.* The long, low lines of the *Messerschmitt* enable you to corner at speed with complete safety, for the car clings to the road with unbelievable tenacity.

3 *Easy Parking.* The streamlined, compact shape makes parking child's play—even in the smallest space.

4 *Easy Entering.* The swing-away roof allows you to enter the car standing upright. This is unique to the *Messerschmitt*.

5 *Perfect View.* You have uninterrupted, perfect all-round visibility in your *Messerschmitt*, making manoeuvring simple and safe.

6 The *Messerschmitt* is available in various attractive colours with superb finish, as Sports saloon, Convertible or as Cabrio with Folding Canvas Hood.

See your dealer now! There are Messerschmitt dealers everywhere, fully equipped for servicing and spares. If you wish to read what independent experts have written about the Messerschmitt, send a postcard, giving your name and address to:

SOLE CONCESSIONAIRES FOR THE UNITED KINGDOM

CABIN SCOOTERS (Assemblies) LTD.,

Dept. 23, 80 George Street, London W.1
HUNter 0609

seater comfort in the NEW

MESSERSCHMITT CABIN SCOOTER

KR 200

CAR COMFORT MOTOR CYCLE ECONOMY

A new 3-seater Messerschmitt with all the joy, comfort and economy of the KR 175 plus many new features! New steering, new lighting, new controls, new panoramic windscreen, all provide even greater comfort, even greater driving ease.

200 cc 2-stroke engine gives 62 m.p.h. 100 m.p.g. hand-operated 4 speed gearbox. Electrically operated reverse gear. Foot operated clutch, brake, accelerator and dipping switch. Electric self-starter and windscreen wiper. Double-dip twin headlamps, flashing indicators. All-steel body. Folding rear seat. Luggage locker. Foam rubber upholstery.

Basic Price	£294 . 17s . 1d
Purchase Tax	£60 . 16s . 3d
Total	£355 . 13s . 4d

Deposit £119 . 13s . 4d 24 MONTHLY PAYMENTS of £11 . 16s . 0d

MESSERSCHMITT

Sportwagen

4-rädrig mit der Leistung eines Sportwagens

Masse:	Länge 2860 mm, Breite 1270 mm, Höhe 1150 mm, Bodenfreiheit 150 mm, Spurweite vorn 1110 mm, Spurweite hinten 1040 mm, Radstand 1885 mm, Eigengewicht ca. 330 kg.
Rahmen:	Verwindungsfreier Stahlrohrrahmen mit vollkommen geschlossener Bodenwanne.
Lenkung:	Sehr direkte, verschleissfeste Achsschenkellenkung mit Spurstangen in Kunststofflagern. Griffsicherer, handgerechter Lenker mit Signalknopf.
Federung:	Einzelradaufhängung mit Progressivfederung.
Bremsen:	Grossbemessene hydraulische Bremsen 180 mm ϕ.
Karosserie:	Ganzstahlkarosserie mit wetterfester, gebrannter Kunstharzlackierung.
Innenraum:	Hintereinanderliegende, schaumgummigepolsterte Sitze. Vorn Bereitschaftssitz, wie er speziell für das Weltrekordfahrzeug entwickelt wurde. Rückwärtige Sitzbank über die ganze Breite des Fahrzeuges, geteilt, hochklappbar. Hinterer Sitzraum Platz für 1 Erwachsenen und ein Kind.
Bereifung:	4,40 × 10.
Scheinwerfer:	130 mm ϕ in modernster Ausführung mit asymmetrischem Licht.

	Mit 19,5-PS-Motor	Mit 24,5-PS-Motor
Leistungsgewicht:	16,9 kg/PS	13,5 kg/PS
Geschwindigkeit:	Spitze ca. 130 km/h	Spitze ca. 140 km/h

Beschleunigung:	Von 0—80 km/h unter 14 Sekunden.
Motor:	500 ccm-Zweizylinder-Zweitakt-Motor Fichtel & Sachs, mit Gebläsekühlung, Drei-Stufen-Vergaser, 4 Gänge, Rückwärtsgang.
Preis:	Je nach Ausführung ca. Fr. 4'500.—.
Liefermöglichkeit:	Spätsommer 1958.

Importeur und Generalvertreter für die Schweiz und Liechtenstein:

AUTOMOBILWERKE FRANZ A. G. ZÜRICH

Badenerstrasse 313   Telefon 051 - 52 33 44

Vertreter in der ganzen Schweiz

ORGANIZED by the Oxford University Motor Drivers' Club, the Targa Rusticana series has enjoyed increasing popularity and this year attracted a fully subscribed entry of exceptional quality.

All credit must go to John Brown, clerk of the course, and his assistants, for their choice of a novel and tough route on sheets 128 and 141, ably backed by efficient marshals and results team.

The first of the 70 entries started from Winforton, near Hereford, at 9 p.m. on their sealed watch to tackle three 90-mile loops, each composed of very short, tight sections at a true 30 m.p.h. average.

Well sited controls, seldom unmanned, with plenty of white road motoring in between, soon set the pattern, and no one remained clean by the time Llan-

heavy penalties indeed; even Bill Bengry's VW had to be lifted from a cosy hedge bottom between controls 32 and 33 near Bwlch-y-Ffrid, costing him 10 minutes, a bent front wing and injured pride.

Larger cars, such as Derek Astle's recently acquired ex-works Healey 3000, must have required delicate handling on the very narrow roads.

On this middle loop, too, the Piper/Coombes Messerschmitt had retired with suspected engine failure, and the M.G.A Twin-Cam used by the organizers to head the competitors and ensure that all gates were open (a wise move with the increasing popularity of safety belts) had collected a holed sump on one of the many humps encountered *en route*.

Prior to the final loop, 15 minutes was

previous two loops, the final one to breakfast at the Craven Arms still required the utmost effort from driver and navigator alike, and one again via Llanbister to Hartseas and Baillie Hill to Clun Forest, consistently varying in length from two minutes to 11, continued to reward the efforts of the route-setters.

The results team quickly announced that Julian Easten/Graham Robson, in John Sprinzel's rapid Mini, with prototype Weber installation, secured a narrow victory over Don Grimshaw/Brian Melia, 2.2-litre TR3A.

One thing is very certain, the Oxford University organizing team have set a target for 1961 restricted rallies, and even nationals, that will be very hard to beat.

JULIAN EASTEN.

Results

1, J. Easten/G. Robson (Mini-Minor), 8 mins. lost; 2, D. Grimshaw/B. Melia (2.2 TR3A), 9; 3, B. Harper/R. Crellin (Morgan Plus 4), 11; 4, R. McBride/D. Barrow (Anglia 105E), 13; 5, A. Bengry/D. Skeffington (VW), 22; 6, R. Fidler/P. Lichtensteiger (Anglia 105E), 34.

Team Award: Kowldale Car Club (D. Grimshaw, R. Fidler, B. Harper).

THE "TARGA RUSTICANA"

An Impeccably Organized O.U.M.D.C. Event

drindod Wells was reached at the end of the first loop.

Seasoned competitors, including Bill Bengry/David Skeffington (VW), Don Grimshaw/Brian Melia (TR3A) and Ken Piper/Ken Coombes (Messerschmitt) had all accumulated the odd lost minute on the tighter sections, such as those between controls 13 and 14 (Glascwm via Rhulen to Cregrina) and 16 and 17 (Gilwern Hill), all on narrow, twisting roads, made more difficult by mist patches of varying density, and noticeable gaps had appeared in the column even at this early stage.

After a 45-minute halt at the Metropole Hotel for fuel and refreshments, the last 15 minutes of which was set aside for plotting the second loop, a cheerfully confident clerk of the course despatched cars on their way to a series of sections between Llanwrthyl and Elan Valley—Llangurig and Glyhaffren Bridge and near to Trefeglwys—with controls placed at intervals of 3, 4 and 6 minutes respectively.

Any wrong-slotting, or momentary relaxation by crew members, incurred

allowed for plotting and refuelling at Dolwen, by which time most crews were running very late, incurring failed sections on top of time penalties.

Further de-ditching operations had also been reported, including the rapidly conducted 122S Volvo of Jenny Law, the Arthur Dryden A40 and the Herald of last year's Targa winner Richard Russell.

Perhaps a little easier than the

ABOVE: Ken Piper's Messerschmitt was forced to retire with engine trouble.

RIGHT: Bill Bengry took fifth place, despite an excursion to a ditch which cost him 10 minutes.

Renseignements intéressants sur le

Scooter à cabine

MESSERSCHMITT

Le scooter à cabine MESSERSCHMITT est un véhicule d'un genre tout à fait nouveau; il abrite ses passagers 100 % contre les intempéries. Ce scooter fermé est le moyen de transport idéal pour tous ceux qui tiennent à se motoriser d'une manière économique, pour qui le facteur temps et des vêtements toujours impeccables jouent un rôle primordial, et que le scooter ordinaire ne satisfait pas ou ne satisfait plus.

Les indications ci-dessous sont destinées à donner aux personnes qui s'intéressent au scooter à cabine MESSERSCHMITT une idée précise de ce nouveau véhicule.

Permis de conduire

Toute personne qui possède le permis de conduire pour véhicules de la catégorie H (tricars) peut conduire le scooter à cabine MESSERSCHMITT. Les personnes sans ce permis, mais possédant le permis pour la catégorie A (automobiles légères), ou F (scooter, motos), ou G (motos avec side-cars), n'ont plus qu'à se soumettre, dans le canton de Genève par exemple, à un simple **examen pratique** d'environ 20 à 30 minutes. Une fois en possession du permis pour la catégorie H (tricars), les personnes désirant acquérir plus tard le permis pour la catégorie A (automobiles) n'ont plus qu'à subir un examen pratique; elles sont dispensées de l'épreuve théorique.

Impôt et assurance

A ce point de vue là également, le scooter à cabine MESSERSCHMITT est le véhicule fermé le plus économique. Ci-après, nous vous donnons un tableau comparatif des montants auxquels s'élèvent ces primes d'assurance et impôts pour le canton de Genève:

	plus petite automobile	scooter à cabine MESSERSCHMITT	plus petite moto plus petit scooter
Taxe annuelle pour le canton de Genève, par ex.	110.—	80.—	25.—
Assurance resp. civile obligatoire tarif valable pour toute la Suisse	237.—	115.—	95.—
Total	347.—	195.—	120.—

Le scooter à cabine MESSERSCHMITT ne revient donc qu'à Fr. 75.— de plus que le scooter ordinaire de 125 cm³ avec siège arrière. Pour une moto entre 125 et 250 cm³ avec siège arrière, la prime d'assurance s'élève même à Fr. 185.—.

Pour le canton de Genève, le scooter à cabine MESSERSCHMITT revient donc à Fr. 30.— meilleur marché qu'un scooter avec cylindrée identique.

SUVA, accidents non professionnels et assurance accidents personnelle.

La Caisse Nationale Suisse d'Assurance SUVA dédommage les employés d'une entreprise lui étant subordonnée pour les frais pharmaceutiques, de médecin, et pour le salaire, si les dits employés se cassent par exemple la jambe à ski ou lors d'un tour en automobile. Cependant, SUVA refuse de verser la moindre indemnité lorsqu'il s'agit d'un accident de moto. **Elle couvre donc les accidents non professionnels à l'exception des accidents de moto.** SUVA, et par conséquent toutes les assurances privées, couvre le risque pour le scooter à cabine MESSERSCHMITT, car elle le place au même niveau qu'une voiture. Ceci prouve combien elle estime petit le risque d'accident avec le scooter à cabine. Pour les personnes assurées auprès de la SUVA, il n'est donc pas nécessaire de contracter une assurance accidents spéciale, ce qui reviendrait par année à Fr. 100.— au moins.

Importateur et agent général pour la Suisse et le Liechtenstein

AUTOMOBILWERKE FRANZ S.A., ZURICH

Badenerstrasse 313   Téléphone (051) 52 33 44

et environ 80 agents dans toute la Suisse, ainsi que des représentations générales dans la plupart des pays d'Europe et d'outre-mer.

BUBBLE CARS

Post-war Britain had a bit of a problem as far as cars were concerned. In the first place there was a distinct lack of money, which didn't help much. The other problem was that even for those with money there just weren't any cars around. All that manufacturers could offer was a waiting list, and a waiting list for pretty uninspiring cars at that. The situation was ripe for some enterprising manufacturer—some new manufacturer, even—to come along and offer cheap motoring. And sure enough, they came in their droves. After all, bandwagons exist for jumping on.

From the very beginning of motoring there have been three-wheelers in this country. Between the first and second world wars the Morgan emerged as the epitome of the three-wheeler. It was little more than a motorcycle with an extra wheel, it was fast, noisy and uncomfortable, and yet people loved it. In fact the name Morgan (I'm reliably informed) came to be used as the general name for the species in much the same way as Hoover later became the household word for vacuum cleaners. The Morgan and its ilk were sporty enough, but not really in the running when it came to everyday transport for the masses. It was a young person's car, and an impecunious one's at that.

After the war there emerged a new type of three-wheeler. It was aimed at the masses, and pretended to be a scaled-down family saloon—it had little to offer but its cheapness, and people with little or no discrimination forgot about the missing wheel and went for its low road tax, its good petrol consumption figure, forgetting that what it made up for in monetary terms it more than compensated for with a complete lack of charm.

Credit for the first of this new breed must go to a Preston company, Sharps Commercials, who produced the Bond Minicar in 1948. Throughout the fifties

and early sixties it was followed by a plethora of other makes—Frisky, Coronet, Reliant, Berkeley, Noble, Scootacar, AC Petite, to name but a few—all of whom hoped to emulate the success story of another distinct type of three wheeler, the bubblecar.

The history of the bubblecar is an interesting one, and one which makes it stand out as a very different type of vehicle from its British 'counterparts'. The most obvious distinction was that it didn't pretend to be anything but itself, and while it maybe left a lot to be desired in the aesthetics department, there was no denying the individual approach and the subtle difference between the European and British cars.

On the Continent, you see, money was even shorter than in this country. Surprise surprise. Nobody knew at the time that this economic state would eventually be used to their advantage, so meanwhile the factories once devoted to wartime production looked around for something to take its place. War-torn Italy sired the Vespa scooter and later the Lambretta, and as sales soared the prospect of unemployment and poverty began to mean less for a select few. In Germany, a few years later, one-time manufacturers of aircraft, or at least companies operating under the old names, began to make three-wheelers.

A lot of the principles of the motor scooter were very apparent, but on face value they represented a new type of vehicle.

The three German-built bubblecars that reached this country in those early years were so utterly different from the Bond that they caught on immediately. There was no pretence—they weren't just conventional cars, albeit small ones; they were more sophisticated than that. There was no need for slab-sided styling to disguise a basically cramped motor car. They were different, fun to drive, handy in traffic—they represented a new approach. Almost certainly they sold initially on their low price, but in years to come they were to be looked on favourably for different reasons.

The 174cc Messerschmitt Kabinenroller was first exhibited at the Geneva Show in March 1953. Kabinenroller, incidentally, means nothing more than cabin scooter, the reasons for which name you can decide for yourselves. The British motorcycling press was prompted to remark shortly afterwards that '. . . it created something of a furore. Nothing quite like it had been seen before, and its aircraft ancestry was apparent in every line. . .' It went on to announce that they were being produced at the rate of 200 a week, and that arrangements were in hand to import the new car into Britain.

Certainly it resembled the cockpit section of the 109, and the name itself left little doubt as to its manufacturer. But the story goes back a little further than that. In fact, the Messerschmitt as it's known today was invented by one Fritz Fend, whose 50cc Combo Scooter had been based on . . . wait for it . . . an invalid carriage. The link between both types of vehicles was to recur in this country much later, but for the time being the link was a tenuous one, and one that nobody made much mention of. Fend's Combo scooter, familiarly known as the Strassenflitzer, went through a number of engine and styling changes, and sometime in the summer of 1952 began to resemble the 'schmitt as we know it. At this stage Fritz Fend contacted Willy Messerschmitt, who just happened to be looking for something to manufacture. The nasty, mean Allied forces, you see, weren't too keen for Willy to carry on producing fighter planes. . .

The Sachs-engined 175 was long, low and narrow. There was room for two, sitting tandem-style, and the interior was more like a tiny aeroplane than a car. There was no steering wheel, for a start; instead it had a pair of handlebars with a twistgrip throttle on the left. Clutch and brake pedals were more conventional, being foot operated, and instrumentation was necessarily sparse. Gear lever was there by the driver's right hand, and driven properly the little 9bhp two-stroke machine was quite nippy, handled well, and returned 85 to the gallon. It would cruise at 45 to 50mph, road tax was £5 a year and at £335 including import duty and purchase tax, it was thought by quite a few Britons to be worth the money.

Production lasted for two years, and in 1955 the KR200 was introduced. It appeared in four basic types—dome top (saloon), cabriolet (soft top), roadster (with a different hood arrangement) and the Sport, which was never imported into this country. The engine by this time was a 9.7bhp 191cc two-stroke Sachs, and though basically the same as the 175, the car included such innovations as a foot-operated throttle. The bigger Sachs engine thrived on high rpm, and while the top speed went up, the consumption went down. Price was very much the same as the 175, and something like six and a half thousand were imported into Britain, compared to the two or three hundred of the 175.

Between 1955 and 1958 the KR200 had Siba electrics, and in '59 went over to Bosch. The system was the same, but various detail changes divided the species into two distinct types. Nearly all the American 'schmitts, incidentally, seem to be of the earlier type. The KR200 had an ingenious electrical reverse system which neither of the other bubblecars had. Reverse gear was engaged by stopping the engine and then starting it in the opposite direction. Although irritatingly timewasting when the car was being manoeuvred, it gave four reverse gears for the daring, and presented about the only way a 'schmitt could be turned over.

The doyen of the Messerschmitt world appeared in 1958, and lasted through to 1962. It was called the Tiger, or Tg 500, and something like 500 were produced, although only 20 ever made their way to this sceptred isle. The Tiger, with its four wheels, 500cc twin engine and redesigned rear end is today every Messerschmitt fanatic's ambition—the ultimate Messerschmitt, the fastest bubblecar in the world. In 1956 a streamlined KR200 averaged 65mph round the Hockenheim circuit, in a non-stop 24-hour record-breaking run. To do it, it reached 90mph at times.

While the Tiger was undoubtedly faster —the production model was good for 90mph—it was always something of an experimental car and never really became reliable. Supposedly designed by Fritz Fend, an air of mystery surrounds the car to this day. The engine, for instance —was it made by Sachs? It's said that

Opposite page: three prime examples— '62 Isetta, '59 KR200 'saloon' and '59 Heinkel. Bottom pic shows sparse interior of the 175 'schmitt exhibited at the Geneva Show, March '53. All pics courtesy MotorCycle, London.
This page, above: Frau Ilse Hess and son Wolf sitting in an early 'schmitt, presented by Messerschmitt himself. Picture: Fox Photos. Below: two totally tough Tigers, immaculate white one owned by Conrad Moore of Denham, Bucks. Pics courtesy Messerschmitt Owner's Club (London and South Eastern).

Sachs made it, somewhat reluctantly, but never put their name to it. FMR, the company formed by Fend, is supposed to have offered the Tiger to the German police, but were never taken up on it and began to lose interest.

There's no doubt about it, the Tiger went like a bomb. Acceleration was terrific, it handled marvellously, manoeuvrability in traffic was second to none, but every so often they had to be taken to pieces and put back together again—a true enthusiast's car. The 20 that came into this country were imported, not by the concessionnaires, but by Testwoods in Southampton. It was inevitable that they should be raced, and with a recorded 110mph flat out they did very well, but suffered as a result. Today there are still 13 known examples in this country, although only half a dozen or so are actually on the road. There's little chance of finding one tucked away in a barn somewhere, as many more than this have been cannibalised to get to this figure.

In 1964 the FMR factory at Regensberg suffered a disastrous fire. They lost everything, and decided to call it a day. By this time the Minivan had appeared in England, and was selling for £390 or so against the KR200's £350, so the advantage of a Messerschmitt costwise was beginning to fall off.

As soon as Fend pulled out, second-hand prices fell with a bang. You could pick up a 'schmitt for £5, and a lot of people bought them for a laugh, intending to run them into the ground. But it was at this time that the two Messerschmitt owners' clubs picked up a big influx of members, many of whom went on to be dedicated 'schmitt drivers and owners, interested in the vehicle for its own sake.

A Messerschmitt owners' club was formed at the same time as the 'schmitt appeared in this country, and was followed soon after by another club. Both clubs— Messerschmitt Owners' Club (London and South Eastern), the original club, and The Messerschmitt Owners' Club of Great Britain—are alive and well today, and together with the Frankfurt Offenbach Karofreunde belong to the International

Federation of Messerschmitt Clubs. Les Tilbury, Chairman of the MOC and editor of Kabinews (incorporating Messerschmitt News and Karo-Nachrichten), the magazine shared by all three clubs, supplied much of the information given here, and can be reached at Eleanors Cottage, Little Braxted, Witham, Essex. This far-reaching club is doing much to promote interest in 'schmitts in this country, and the way they have the spares situation tied up and their very serious attitude to a much-maligned and near-forgotten side of motoring is nothing short of praiseworthy. It's nice to see that the interest's been retained, more so since both Heinkel/Trojan and Isetta owners are on their own and nothing short of quick formation of a club will keep these interesting vehicles on the road for very much longer.

The very first KR200 ever produced was given to the American celebrity Vic Hyde. Thousands of Messerschmitts went to the States, and today the club has over 70 members out there. It's doubtful if the Tiger was ever imported by the concessionnaires—what examples exist (at least five are known) were probably imported privately. Many of the KR200s—and 175s, come to that—appear to have been bought as gimmicks. One club member in Los Angeles, for instance, recently bought 15 from Bob Hope, while another fully restored 'schmitt was sold a couple of years ago to a wealthy Jewish gentleman for $1400—as a talking piece in his living room! Let's hope it never happens in this country, for it will almost certainly mark the beginning of the end for the devout and almost fanatically enthusiastic members of the Messerschmitt Owners' Club.

The nearest a British manufacturer ever got to the Messerschmitt was the Leeds-built Scootacar, which appeared in the early sixties. Starting with a 200cc engine, it went up to 324cc which with a glass fibre body made for quite a turn of speed. But it was really ugly, and just not in the same category as the 'schmitt. It was like driving a telephone kiosk, and though there are still a few around today, they've never been classed as bubblecars so I won't say much more. Production stopped around 1965, and the only reason they ever sold at all was the fact that they were British, not to mention £50 or so cheaper than a 'schmitt.

Of the three makes of bubblecar—Messerschmitt, Heinkel (later the Trojan) and the Isetta, the 'schmitt was the thinnest, the Isetta the fattest (so far that it nearly falls into the mini- as opposed to micro-car classification), with the Heinkel halfway between the two, rounded at the front and tapering towards the rear. Personally, it is my favourite from a looks point of view, though the Messerschmitt leaves it standing on most scores. As far as the history goes it has one or two black marks, which I'll be coming to in a minute.

The Heinkel, like the Messerschmitt, relies on an erstwhile aeroplane manufacturer for its name and to some degree its reputation. It started life much later than the 'schmitt, and was built in four-wheeled form by Heinkel in Stuttgart between 1956 and 1957. The home market wasn't all that Heinkel had hoped it would be, so they looked round for someone else who could make it.

At the time Heinkel were making a scooter called the Tourist, reckoned by some to be the Rolls-Royce of motor-

Above: two lovely schiny 'schmitts, a '59 KR200 Roadster at the top, cabriolet version with period piece below. Pics: MotorCycle, London

scooters. It was being made under licence in Dublin by a firm called Lincoln and Nolan, who also assembled Austin cars. One of the directors was a German, or so the story goes, and Heinkel approached him to see if Lincoln and Nolan could take over the bubblecar. Production, in three-wheeled form, was started, but Austin began to moan and something had to be done.

This was at the time that railways were being closed down in Ireland, and the Irish Government was faced with disused railway works and a large work force on the dole. By a strange coincidence another of the directors of Lincoln and Nolan happened to be a Finance Minister or something so the Irish Government was persuaded to take over the manufacture of the Heinkel. Dundalk Engineering was the result, with a firm of concessionnaires set up by Lincoln and Dolan called International Sales, to bring the car into England.

Unfortunately, Irish railway workers and a bubblecar production line were just not compatible; the press tools and dies were neglected, pressings came out an odd shape and the general quality of the Heinkel simply wasn't what it should have been. The engine, however was still made by Heinkel, so that at least was OK. And it continued to be made by Heinkel, right through to the end.

Eventually though, like all good things this state of affairs had to come to an end and in 1960 the Purley Way, Croydon, firm of Trojan bought up the manufacturing rights. Trojan carried on producing the car, now renamed, until 1965 or 1966 when the state of the press tools, and the simple laws of supply and demand meant only one thing—they couldn't afford to produce it any more.

Produced first of all with almost the same 175cc engine of the scooter, the Heinkel was given a 198cc engine during the time it was produced in Ireland. It is this engine which is the most familiar, and apart from a left-hand to right-hand-drive option offered by Trojan, the car remained virtually unchanged over the years, except after 1960 it was made a little better, of course.

Like the Isetta, the Heinkel/Trojan had a large door at the front, good all-round vision (a characteristic of all three) and a sunshine roof. The four-stroke 10bhp 198cc engine was good for 55mph on the flat, and when driven hard still produced over 70 to the gallon. Only one reverse gear this time, but even so the subject of much discussion 10 years ago. In August 1963 the driving-licence group anomalies were sorted out and it was decreed that anyone with a full motor cycle licence was entitled to drive any type of three-wheeler, with or without a reverse gear.

In 1963 the Trojan cost £370 with right hand drive, £10 less for a left-hooker. The steering wheel remained where it was when the door was opened, unlike the Isetta which had a clever arrangement whereby the steering wheel moved out of the way. Simple gearshift and handbrake were on the left, and instruments were, well, basic. The eventual demise of the Trojan is—like the Messerschmitt—directly attributable to the cheapo cheapo Minivan, and with it went another little bit of history. Trojan kept the bits and pieces for a number of years, and when they couldn't find anyone to take the lot

1960 Heinkel Cruiser, price £403 at the time. Note fixed steering column, opening side windows. (MotorCycle pic)

appearance it was quite similar to the Heinkel, except the back was more rounded.

1958 saw the introduction of the 295cc Isetta Plus, a three-wheeler designed to take advantage of the cheap road tax. By now it was made by Isetta of Great Britain, formed in 1956. Isetta GB had been formed to import cars, 1000 had come into the country, and they then decided to start production here, beginning with the four-wheel version. Externally there was little difference between the old and new models—there were obviously one or two but you'd only accuse me of pendantry if I began to point them out. Price by now was down to £365—definitely a step in the right direction.

A right-hand-drive version of the Isetta Plus appeared in 1960. Again detail differences only of interest to the purist, but the price had come down to £359. There wasn't much between all three by this stage on the price, so it merely remained to see which proved the most popular. Well, large scale production of the Isetta stopped in 1962, and had finished altogether by 1964. Four wheels were proving to be infinitely more desirable than three, and with vans minus the burden of purchase tax the bubblecar became a thing of the past.

Wouldn't it be nice to get the bubblecar thing going again? There are usually half a dozen or so in Exchange and Mart, and they're cheap enough. Who's going to start the ball rolling then. Who's going to KUSTOMISE one for next year's show? It's not as silly as it sounds. Let's see now, there are enough spares people around if you look for them; I've even had a look round myself and found some easily. Messerschmitts are pretty strong on the ground already; how about a bit of care for the old Heinkel and Isetta, eh?

Could just be an investment, you know. Until a few weeks ago when I started to look into bubblecars (got a crick in the neck in the process) I hadn't given them much thought. But when you get down to it, there's a lot more than meets the eye. Anyone got one they don't want? I'll take it off your hands. **CG**

Top: rhd Trojan—45mph cruising speed, 55mph flat out, 95mpg. Above: rear view, '58 Heinkel. Both pics: MotorCycle

Above: 1962 rhd Isetta. Below: Fox Photos pic dated 1 October '59 shows canvassing Wandsworth Labour candidate.

off their hands, got rid of them in dribs and drabs as and when they could. Trojan themselves (who recently went into liquidation, though this has got nothing to do with it) no longer have any spares or interest in the car, though the list at the end gives a couple of stockists.

Last, but by no means least, to use the most suitable cliché, we have the Isetta. Designed by Iso of Milan, hence the diminutive, it was manufactured under licence in Germany by BMW, the not unknown motorcar and bike firm.

'The BMW Isetta Motocoupe' announced The Autocar of 4 November 1955, 'now for general sale in Great Britain . . . attracted considerable interest at the recent show at Earls Court'. It pointed out that the Isetta was still produced at the time by Iso, and was also being made in France, but it was the BMW car that we had in this country. I was only five at the time so I'll have to take their word for it. The four-stroke 245cc engine put out a maximum of 12bhp, and propelled the little car from 0-30 in 12½sec, with a top speed of 54mph and anything up to 74mpg. List price, including purchase tax of £123, came to a massive £415. It might have been small, but it wasn't that cheap. It was available in left-hand-drive form only and had twin wheels at the back. In

Technical Instruction Nr. 6

NOTE

The hood is oversize and will have to be machined down to correct size.

Directions for fitting KR 200 hood assembly

① Take away any remained glues or dirt from the aluminum framework.

② Place the safety-glass hood so that it fits into the aluminum-alloy framework (picture 1).

③ Mark round the full circumference of the hood by running a scriber along the top edge of the framework (picture 2).

④ Distance between scriber and outside edge should be 9 mm. Oversize to be machined down (picture 3). The hood fits when it rests hard against the inner wall of the framework (picture 4).

⑤ Before mounting, the hood "Bostik" gum should be pressed against the inner wall of the framework (picture 5).

⑥ Place the hood into the framework press rubber tubing into place in the channel of the framework regularly working from the center of the windshield towards left and right with the help of a suitable shape wooden wedge. Any surplus rubber tubing left at the rear should be cut off.

During this procedure hold the hood down into the channel of the framework (picture 6).

①

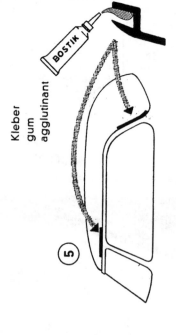

②

richtig
right
propre

falsch
wrong
incorrect

④

Kleber
gum
agglutinant

BOSTIK

⑤

③

abschleifen
smooth
égriser

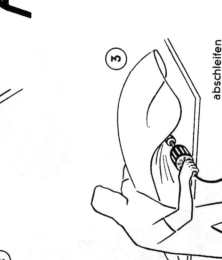

⑥

Gummischlauch eindrücken
compress rubber tube
tryau de caoutchouc im primer

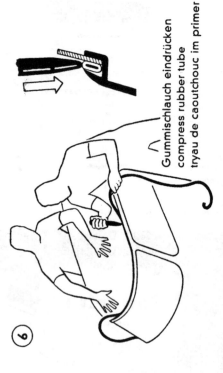

FⓍMⓇR   FAHRZEUG- UND MASCHINENBAU GMBH., REGENSBURG, Lilienthalstraße

Tel. 30773 / 30774 - Fernschr. 065882

Phil Boothroyd Recalls

THE DAY THE TIGER CAME TO ENGLAND

This month I am taking you back more than twenty years to an exciting event in Messerschmitt history which is still talked about by those who witnessed it. With the aid of personal recollections, my own and those of others, I shall attempt to set the scene as it was in 1958 leading up to the arrival of the first Tigers in Britain and the demonstration given by Ing. Fritz Fend in his now legendary 'R-AX 350' at Narborough, Leicestershire.

People often say they can recall precisely where they were and what they were doing when some great international event occurred. Perhaps the announcement of a new type of car, even one as unusual as the Tiger, does not quite fit into that category but in my case it does. If you are old enough you may remember the epidemic of Asian 'flu in the Autumn of 1957. In late September of that year I happened to be in Newcastle upon Tyne and one sunny afternoon felt I had caught the bug. Going into Fenwicks, the big department store in Northumberland Street for copies of Autocar and Motor magazine, I then caught a bus...this was before I even owned a KR200 ....to the large park known as the 'town moor' to get some fresh air. Feeling very rough I sat down on a park bench and began to look at the reports of the '57 Frankfurt Motor Show .... and there it was, a picture of a new Messerschmitt with four wheels. The captions, in Autocar particularly, were written in condescending terms but the effect was exciting and therapeutic to such an extent that four hours later all symptoms of the 'flu had disappeared.

Following that announcement the big talking point at all club meetings in those days usually began with the questions "Are you going to buy a four wheeler?" It seems strange today but that was how we always referred to the Tiger and in the absence of any definite information about the price and performance the discussion usually centred on whether we could afford to go up in the £12.50 annual taxation class from the £5 we were then paying on our cabin scooters.

FMR had encountered legal obstacles over their initial use of the 'Tiger' name similar to those which had already required the company to abandon the famous bird motif and which were about to compel them to modify the 'three rings' version of the FMR emblem. Interesting though the reasons are they need not detain us and in any event they failed to prevent the name coming into general use.

On Monday 5th May 1958 at a ceremony and reception held in brilliant sunshine at the Regensburg plant the car was officially launched and demonstrated to the Press and dealers under the compromise designation 'TG 500'. Incidentally those who know the factory will have noticed the large open space that borders it to the west just beyond the rail spur which connects with the main Nürnberg Ingolstadt line. This was formerly in use as a horse racing track and until recently I believed it to have been the venue for the demonstration. However, in the light of recent research it seems more likely to have been held on the former factory aerodrome which lay immedi-

ately to the north and which can be identified by anyone interested by the large Siemens plant which today occupies part of it.

No news of this major event filtered through to Britain at the time but our assiduous general secretary, Ernst Hartmann, was in frequent contact with FMR and keen interest and speculation developed throughout the summer. Obviously an announcement was imminent. It came in the form of a club circular to all members informing us that (instead of the home servicing session we had been expecting) on Sunday 28th September at Narborough we would be meeting the inventor and designer of our cars, Ing. Fritz Fend, who would be driving his own TG 500. That date is embedded in my memory as it was itself the 20th anniversary of the ill-fated Munich Agreement.

Naturally what had been originally planned as a routine Midland Area gathering was about to take on the character of a national event with the London and South Eastern Area particularly well represented. Now two of their keenest activists were to be the first to sight the Tiger.

Ken Forgan and Fred Borst had set out early from North London in their KR200s and were ready for a coffee stop. Pulling into a lay-by on the A5 they had just opened their flasks when....ZZZZEEEEEIP....and they looked up just in time to see the aerosilver Fend Tiger heading north at full bore. A quick glance at each other was enough. Gulping down their coffee Ken and Fred ran back to their

'schmitts so as not to be late for the fun.

Who was responsible for this inspired and brilliant coup? The credit belongs to two people, Ernst Hartmann and Fred Wharton, managing director of Cabin Scooters (Assemblies) Ltd, the company which from late 1956 really put Messerschmitts on the map in Britain after two others had tried and failed.

It came about like this. Earlier in 1958 and partly with the aim of launching the Tiger, Mr Wharton had engaged the London firm Hurd International Public Relations to give the professional touch to Messerschmitt publicity in Britain. In discussion with them and following the successful Regensburg launch in May, the idea was born to introduce the car at a famous racing circuit. Brands Hatch in Kent was selected and detailed preparations made with the Press, Gaumont-British News and celebrated motor sport personalities like John Bolster. The date was fixed for Monday 29 September 1958 with the Fritz Fend and 'RAX 350' as the star performers. Ernst Hartmann immediately asked Fred Wharton to notify Mr Fend about the club meeting arranged for the day before. Although only coming to England for a long weekend of four days, Fritz Fend, in a gesture typical of him, readily agreed to drive up to Narborough from his London hotel to give club members a special preview of the launching and demonstration.

In the summer of 1958 the English weather had been up

to its tricks again. Throughout August and into September there had been heavy rain with hardly a day's break. Sun had been as scarce as fivers on the plate at a church collection. Rain was to mar the official unveiling of the Tiger at Brands Hatch on the Monday afternoon but Sunday 28 September dawned grey and dry and these were the weather conditions as I went over my KR200 with a soft cloth in final preparation for what time has since confirmed as one of the most rewarding days in my 'career' as a Messerschmitt owner.

The 'schmitt was less than six months old and running beautifully as it still does today. Heading east on the A46 for Narborough I would quite often meet up with the Midland Area Captain, Stan Williams, in his red KR200. A keen driver with sporting inclinations, Stan would usually take the lead and after negotiating the villages of Shilton, Wolvey and Sharnford, there is a straight and undulating section of road built on the site of the old Roman Fosse Way over which a well tuned Messerschmitt rewards its driver with that exhilarating sensation not quite matched by any other vehicle of which I know.

There was already a gaggle of 'schmitts outside the block of private garages where our meetings at Narborough were held and with two other KR200s I parked mine on the grass verge. It was obvious from the start that this was to be no ordinary club meeting. By midday there was an air of expectancy and excitement as the predominantly London and Midland accents intermingled. In those days I held no office in the club and could relax and observe the event to the full. For General Secretary, Ernst Hartmann the day was to prove one of great achievement. Taking charge in that faintly military manner so well remembered by all who knew him, he announced briskly that Mr Fend was on his way and expected shortly. With that Ernst climbed into SXV885, his gleaming cream and red 1956 KR200 which outshone all others and accelerated smartly away to lead our distinguished visitor in from the A46.

We heard the Tiger before we saw it. The attention of the whole group was gripped by the throaty note of the big twin-cylinder motor superimposed on which was the now famous whine from the cooling fans. As Fend traversed the modern housing estate which Narborough comprises, all eyes turned in the same direction and were rewarded with a first glimpse of the car which, to this day, has a special fascination for all 'schmitters.

R-AX350 (chassis No.120570) was at that time finished in aero-silver and had a normal dome top (as it again has today) which Fritz Fend later removed for competition work, I believe during the 1959 season. Non-standard fittings included a stone guard running along the lower body panels (see picture) and twin horns fitted externally.

Although this was the personal car of the designer it did not at that time have so many special features as it is sometimes credited with - in particular the motor rumoured to have been specially tuned by Albrecht Mantzel of DKW fame was just that - a rumour. If we had known then what we know now some points would have come in for closer scrutiny. But this car was our first Tiger and we were examining the whole configuration, particularly the features where it differed from the KR200 which are too well known to need any mention today.

Fritz Fend pulled the car up, right among the admiring group of club members now about twenty-five strong. Alighting with him was Fred Wharton, Managing Director of the Importers, then, as now an immaculate figure. Despite his many achievements Fend is not a fluent English-speaker so both Mr Wharton and Ernst Hartmann had to work overtime translating the avalanche of questions. In his replies Mr Fend stressed the low centre of gravity and wide margin of safety he had designed into the Tiger as a car almost impossible to roll. This was not mere talk as he was later to demonstrate.

Ken Forgan, Fred Borst, Stan Williams and the other technically minded members present were impatient for a look at the motor and Fritz Fend obliged by removing the bonnet. It seems surprising today that there were no questions about the origin of the 500L motor but we have to remember that the emergence of the Tiger as a 'classic car' then lay in the distant future and so did questions relating to its genealogy.

No special claims were made for the Tiger's maximum speed but the designer was justifiably confident that the car's handling and cornering characteristics were exceptional even by sports car standards. A technical innovation on which Fritz Fend expounded was the rear suspension layout, in particular the self-compensating toe-in arrangement and its effect as a safety measure when the car was being pushed hard on corners as he anticipated it would be in competition work. This feature, of course, typifies the Fend approach to design - simple and effective. At the time the full significance of what was being explained was partly lost owing to the language difficulty. Years later Mr Fend was to tell Klaus and Bernd Eigenfeld that at that time he could not believe no one had conceived the idea before and so had not applied for a patent. Some day it would be valuable if we could have the system explained in Kabinews, especially now that the same principle is embodied in the 1979 'Car of the Year', the Porsche 928, where it forms part of the 'Weissach axle'.

What then, on that grey September day twenty years ago, was our impression of the man whose ingenuity and foresight had provided us with the cars which have since so remarkably withstood the searching tests of time? The

Messerschmitt is an unusual vehicle and so is the man who designed it. Fritz Fend was the first visitor from Germany I had ever met, which, to younger members accustomed to today's easy and frequent Continental travel, will seem surprising. What I noticed about him was what I can best describe as an 'impressive presence', something often discernible in people of exceptionally outstanding ability and stemming, I believe, from an absolute confidence not attainable by us ordinary mortals with more pedestrian abilities. In a country renowned for such skills, Fritz Fend is most admired as an innovator, particularly in engineering design, finding simple, practical solutions to difficult problems often leading to commercial success. It would take too long to catalogue them all but as an example, the next time you raise the bonnet of your KR200 just take a really close look at the design of the rear frame and suspension.

Does all this mean that Fend is a dry, studious boffin type of chap? Not a bit of it. As he went among our group of 'schmitters, shaking hands with everyone and making the most of his English, the great designer from Bavaria made a profoundly favourable impression with his engaging and warm personality which other members who have since been privileged to meet him have also remarked upon to me. Now we were to witness yet another facet of Fritz Fend's talent as he ran the Tiger motor up in preparation for a breath taking high speed slalom demonstration.

A BRUSH WITH THE "FUZZ"

"The Coppice" Narborough, is part of a pleasant post-war housing estate typical of thousands throughout Britain, quiet and peaceful with no through traffic. Few people were about on this serene Sunday morning of 28 September 1958. The road itself ascends a slight gradient at the top of which is a small traffic island with no outlet. Beyond that are the tall trees from which the name presumably derives.

Back behind the controls of R-AX350, the bonnet in position again, Fritz Fend looked around and decided this would make a suitable demonstration course. As he opened the throttle the factory tuned Tiger motor became turbine smooth. With a clear road ahead the designer did not spare the acceleration, earning admiring glances from the crowd of club members. We were about to learn the meaning of the word 'slalom'.

As Fend reached the island, instead of entering it on the left he instinctively followed continental practice and swung the aero-silver Tiger to the right. On each subsequent run he did the same and this was to have repercussions later on. The object of the demonstration was to substantiate the claim he had just made that the car would not easily roll. Beginning the downhill run on the bone-dry road, Fend surged to about 40 mph and began to

swing from lock to lock, throwing the car in a series of near broadsides of such violence that the white-walled Metzeler tyres were all but wrenched from the rims. Yet incredibly R-AX350 remained stable and never even lifted a wheel, so tenacious was the Tiger's road holding. Quite clearly we were witnessing a performance of consummate virtuosity.

Setting up the car for its second run, the designer now had the measure of the narrow road and driving with tremendous panache, put on an even more astonishing performance which drew spontaneous applause from the incredulous onlookers. By now these had been joined by local residents out rubbernecking at their garden gates. Messerschmitts had long been an everyday sight in this neck of the woods - but this was something else. It may be that 'schmitt activities in the area had been too much for some of the locals and this display was the last straw, for events began to move swiftly. A perspective observer might just have noticed a grim-faced woman clutching her garden gate as though her life depended on it. She was to play a decisive role in what followed.

Fritz Fend now came in for his final slalom run. Demonstrating truly superb control he pushed the Tiger to within a few degrees of a full broadside on each left and right swing. To this day I cannot comprehend how the tyres remained on the rims. The entire group were potential Tiger owners after this gripping performance yet as he swung back the hood the great designer appeared unruffled and relaxed as though he had just come in from a Sunday morning spin. There was a renewed burst of questions and I can still remember the total silence which followed Fred Wharton's reply to the late Esca Mansford's question " how much will the Tiger cost?" Mr Wharton explained that the total price of £651 included no less than £216 Purchase Tax plus Import Duty of £75. The KR200 was then available at £325 and its price was much less inflated by purchase tax which was levied at a lower rate on three-wheelers.

To enable younger members to relate these figures to today's money values, average gross weekly earnings in Britain in October 1958 were £12.83 for male manual workers and £6.70 for women. A Tiger, therefore, cost approximately one year's gross earnings for a man and, ignoring all other factors and to give some indication of whether you would have winced at £651, today's average figures are £79 for men and £49 for women. Although not in the same street as the Tiger, an Austin-Healey 'frog-eyed' Sprite could then be bought for £687 new and a Volkswagen '1200' for £654. The Mini would be introduced eleven months later at only £497.

We were recovering from the shock news of the Tiger's price when an elderly grey Rover 12, vintage 1934, lumbered incongruously into the circle of shiny 'schmitts. The temperature seemed to drop by several degrees - the

'Fuzz' had arrived. As they disembarked from the Rover - this was before the days of Panda cars - it was difficult to tell the rank of the elder of the two police officers as he was in shirt sleeves and wore no hat - then as now, an unusual sight in Britain. Almost certainly he was a sergeant and although his shrewd face creased with a flicker of a smile as he recognised Ernst Hartmann, there was no mistaking the traditional "What's all this about?" approach. Much less sure of himself, his assistant, a constable, was a match for Graham Moffatt who played 'Fatty Arbuckle' in those old Will Hay films. As though about to witness the coup de grace, one or two non-schmitt bystanders including the grim-faced woman, moved in closer. "We've had a complaint that one of these vehicles was being driven dangerously" began the Sergeant as the drama unfolded. Ernst Hartmann was a man of many qualities but diplomacy was not his strong point and instead of trying to damp down the situation he over-reacted. At the same time the police contingent caught on to the fact that this was no ordinary enquiry. Perhaps they had noticed the Tiger's tail with its 'D' - then a novel sight in Britain - or maybe they heard a few words spoken in a strange foreign language.

Later on Ernst was to tell me that as he was regarded as a 'foreigner' the local police were in the habit of calling him in whenever any kind of translation was needed. Only that week a Ukrainian had been found wondering in the fields near Narborough and as usual Ernst had been sent for and told to "get something out of him". Now this relationship, which was useful to the Sergeant, began to exert its influence as he shrewdly realised it would be jeopardised unless he could find an easy way out.

"Are you in charge of this vehicle sir?" The voice lacked conviction and Fritz Fend, who is nobody's fool, replied deliberately in German while our two translators remained 'stumm'. You could see the two officers' minds working as they grappled with the dilemma. 'Fatty Arbuckle' contributed nothing and was clearly out of his depth. A lost dog or a stolen bicycle, difficulties with the local people, these were problems he could handle. But Fritz Fend and Fred Wharton were not local people and a false move might very well provoke an 'international incident'. Yet if nothing was done there may be a formal complaint to the Chief Constable and that too was an unhappy prospect for the older of the two men with a hard earned pension almost within his grasp.

The Sergeant leaned against the old Rover and scratched his thinning grey hair. An unopened copy of the Sunday Express lay on the faded blue upholstery of the front seat. All the signs were that a quiet Sunday morning in the office had been disturbed and inertia began to exert its influence. Sensing imminent defeat the woman who had brought the complaint suddenly stepped forward, arm outstretched like the Statue of Liberty and pointed accusingly at Fritz Fend. "And he went round the island the wrong way" she denounced querulously, a smirk decorating her otherwise bovine features. This was too much for the British sense of fair play and several of us intervened vigorously on behalf of our distinguished guest who had given up his day and driven so far to give us this preview of the Tiger's official unveiling in Britain. To their credit, the two policemen declined to associate themselves with such mean spirited behaviour and upholding the British bobby's world renowned tradition of fairness, pronounced the incident closed, immediately lumbering off in their ageing Rover. With that, Fritz Fend and Fred Wharton adjourned to the Hartmann's for a well deserved lunch over which plans for the 1959 club visit to Regensburg were discussed.

As we drove home, I believe every one of us was determined to become a Tiger owner but as it turned out I was the only one of that group to realise that ambition. Eight long years elapsed before I was able to acquire one of these rare birds - and that, as the saying goes, is another story.

The bubbles that burst!

Derek Pickard recalls an era of super economy motoring of sorts. Could we see such times again?

FAIRLY TYPICAL OF the average young English lad of the fifties, I was the proud owner of a succession of old cars immediately after passing my driving test in 1956.

A Standard Eight of 1938 vintage deposited its innards on the road from Sleaford to Lincoln one sunny summer's day, so that was replaced by a 1932 Wolseley Hornet Special.

Pounds of white metal later, a 1935 Singer Nine Le Mans took over, soon depleting the meager coffers to the stage that motor scooters became more practical from the financial point of view.

But then, soon after wedded bliss threatened to remove the possibilities of car ownership for a long time to come, I bought my first new "car" — a Trojan 200 three wheel concoction then in vogue and known generally as a "bubblecar."

With over 70 miles to the gallon and the ability to cruise at 50 mph over long, non-stop distances, it was just the thing in those days.

Even the inadvertent filling of the tank one day with diesel fuel failed to stop it! I managed to actually drive almost 50 miles on the stuff, even if the

engine would only run flat out all the time!

After a couple of years the Trojan was traded in on a Mini, but I still remember the little beast with affection.

Having taken little interest in the "bubblecar" movement since that time, I often wondered where it all eventually led, this being the reason why Derek Pickard's article which follows intrigued me so.

The story was inspired by a visit Derek made to the First Annual Micro Car Rally in England last year. Recalling a time so different from the present, it's worth reading.

Paul Harrington

ABOVE: BMW lost much face by being associated with the Isetta, an Italian designed three wheeler with opening front.

BELOW: Quite fast as a three wheeler, but better by far as a four wheeler was the Berkeley

ABOVE: Another front opening three wheeler was the Heinkel, later known as the Trojan, a bubble car owned by the editor at one time.

BELOW: The Nobel, looking rather like a later Mini-Jem.

AS A PRINCIPLE, the midget automobile was perfect; its major failing was in being designed twenty years too soon and being nicknamed the "bubblecar".

The era lasted from the middle fifties to the middle sixties. Around ten years of intriguing designs where engineers sought ways of transporting two people in the smallest and cheapest practical method. If their timing and image making was wrong, they also made one more mistake which gave bubblecars a reputation for unreliability. It took a long time for small engineering firms to fully appreciate the problems of dropping an air cooled motorcycle engine into an enclosed car type bonnet.

On paper they had to be winners; cheap to produce, low cost registration, minimal insurance, incredible petrol economy and still the ability to take two people with the added benefit of easy parking. But that narrow concept in automobiles which tries to settle somewhere in between a car and a motorcycle, has always proven a hazardous business, and just as the first such creation in the nineteen twenties (called the Ner-a-car!) went under, so eventually did the "bubbles", the "versions of bubbles" and the "slightly better than bubbles."

The names included; AC, Bond, Coronet, Frisky, Heinkel, Isetta, Messerschmitt, Powedrive, Reliant, Scottacar, Tourete-Supreme, Vernons, Berkely, Peel, Faithorpe, Goggomobile, Opperman etc etc etc. They almost all had three wheels to be registered in Europe as a motorcycle—sidecar outfit and a small commercially available motorcycle/scooter type engine. Most came from samll factories and while a couple lasted the decade, the rest came and went in a few years.

We caught up with some good examples at the Annual Micro Car Rally in England and phtographed some good examples which illustrate the trends of that era.

MESSERSCHMITT

The unique layout and distinct cockpit styling made this machine stand out.

The passenger sat behind the driver who steered with a handlebar control and the only door was the roof!

Called the KR201, it was powered by a 10 b.h.p. German Sachs engine with a four speed gearbox. Like a few other bubbles the reverse was on the ignition timing, in that being a two stroke engine, it could literally run backwards. That meant four forward gears and four reverse!

Being only eight inches, the little wheels guarranteed a rough ride on all but the smoothest surfaces.

An interesting version was the all too brief TG500 Sports Tiger with a big engine and four wheels. It could top 100mph and had a fair bit of competition success in rallies and the like.

They came in 1958 and had all gone by 1962.

ISETTA

This would have to be the worst product to which BMW gave their name.

It originated as an Italian Iso deign with the Germany company taking it over believing (as many people did at the time) the bubblecar was there to stay.

The 250 motor was enlarged to 300cc which cruised at 55mph and gave 60mpg. But valves broke off, transmissions snapped and all manner of other things went wrong.

This bubble which emerged during 1957, well and truly burst in 1964.

HEINKEL

When introduced it had the German name to match the country of origin, but later models were made in Ireland and called Trojan.

The 175/200cc four stroke scooter engine had four speeds and an enclosed chain drive. 10bhp drove the 630lb car up to 60mph with a possible 85mpg.

It had 10 inch wheels, $5\frac{1}{2}$ inch brakes with rack and pinion steering.

The front bench could take three people at a squeeze.

NOBEL

This is an example of the many short lived three wheelers that came from small companies.

Driven by an unreliable little Excelsior two stroke engine, its only distinguishing feature was the ability to take four people; albeit at a slow rate!

BERKELEY

The most beautiful to look at and the worst to maintain! At styling this car was next to perfect, but it had so many mechani-

Bond's "A" Type, the first of quite a successful line of three wheeler economy vehicles.

Much later than the A type was the Bond Bug which enjoyed a brief period of popularity in economy conscious England.

The smallest "serious" production car would have to have been the 50cc Peel, a single seater device using tiny road wheels.

cal faults that it soon became a sick joke.

The choice of two stroke twins and triples, and later big four stroke twin engines was not the best. They weren't suited to such an application and either seized, blew-up or become difficult to maintain.

Other faults were in the transmission, steering and just about everywhere.

With the 700cc Royal Enfield Constelation sports engine, the 50bhp 7cwt car literally flew up to 100mph and the late Lorenzo Bandini took his Berkeley to many racing victories. His 500cc version easily beat the works Fiat Abarths in Italy.

BOND MARK A

The name Bond was one of the most successful. They came in at the beginning and stayed till the very end; one of the very few who kept going.

This is the car which started it all. Released in 1949 the ugly looking Mark A offered little comfort, and no sophistication. The 125cc Villiers engine drove and steered the front wheel. There was no rear suspension, it had a plastic windscreen and a hand operated wiper.

Bond went through the various models (B, C, D, etc) until the mark G in the middle sixties had a 250cc two stroke twin, four seats and an estate rear end. It could cruise at 50 mph and return 50 mpg.

The engines were never completely reliable and with open chain drive, deserved their bad reputation.

PEEL

You're looking at the world's smallest ever car.

Built in the Isle of Man, the Peel was no bigger than a fibreglass shell around a small chair with a 50cc engine tucked under the seat.

It carried one person to 25 mph with nearly 100mpg. But the small 5 inch wheels gave a very rough ride.

During the two years (1965 to 1966) the car was produced, it sold for under 200 pounds and the design was changed from having one wheel at the front and two at the rear, to two at the front and one at the rear. A factor which couldn't have helped its reputation for mechanical sophistication. □

MESSERSCHMITT Continued from page 83

comfortable with the passenger's legs being fully extended alongside the driver's seat. Luggage space was also well up to standard.

Driving at night in the Messerschmitt was not quite so much fun as the headlamps are rather small. The light they gave, however, was quite adequate for normal cruising speeds. They dipped in normal car fashion with the switch being on the floor just to the left of the clutch pedal, although the light switch was awkwardly placed and difficult to operate.

One of the few snags with the car was the gear-change—it was sloppy and did not select accurately. Sometimes it jumped out of gear and sometimes it would not go in properly. Actually, as the model was second-hand, the fault may have been in the cable which operates the gear-change and some adjustment may have put things right.

There were absolutely no complaints about the reliability or performance of the engine. One of the most amazing features is the reversing system. The electrical starting system spins the engine backwards if the key is inserted to its full extent and an amber warning light comes on. If you are daring enough, it is possible to get into top gear while going backwards—a practice not to be recommended! Actually the choice of ratios in reverse came in very handy while reversing backwards in snow. If the wheel spun in the low gears, the selection of one of the higher ones would usually do the trick. If the car got well and truly stuck in a snowdrift, it was possible to drag it out without much effort as it only weighs 506 lb.

In fact, if we have another winter like the last one, I shall seriously consider buying a Messerschmitt. And to have made me consider that is sufficient proof that this little bubble-car is a really impressive vehicle!—B.C. ●

PROFESSOR WILLY

by Phil Boothroyd

A satisfying aspect of writing in Kabinews is the comment and feed-back from members. In this way we have received definite information in 1978 that 'schmitts were built for a time under licence in Italy and we know a lot more than we did a year ago about the Zeta Sports and the Tiger Jet.

The riddle of the Irish '175a' remains to be solved - but experience has taught me to be optimistic as someone somewhere surely has an answer. So, in sending Christmas wishes to 'schmitters everywhere I also thank those who, from far and wide, have taken the trouble to send in items of information. There is still a lot to be learned about Messerschmitt history and the smallest and apparently unimportant details are just as welcome as a major scoop.

A good example is the number of newspaper cuttings which came in from members on the occasion of the 80th birthday of Professor Messerschmitt in June which was followed so swiftly by the announcement of his death in Munich on 15 September. It is not my intention to dampen the Christmas spirit with an obituary but the old year should not be allowed to fade away without some mention of the life and work of the man whose name our movement carries.

The name 'Messerschmitt' means different things to different people. This was brought home to me some weeks ago when Nicholas Parsons in his television quiz game asked a contestant "What is a Messerschmitt?" Back came the answer "an aeroplane" - but I am still wondering what the reaction would have been if the chap playing this game had insisted it was a car! If he had been a reader of the usually accurate London Daily Telegraph he might well have done so. Their report of 16 September not only stated quite unequivocally "Messerschmitt turned his hand to the design and production of (a) pioneer vehiclethe bubble car" but also backed it up with a picture of an early KR 200. The 'Guardian' was nearer the mark and said he "turned to producing scooter-type cars" which is fair enough.

It is a little known fact that the Professor was once quoted as saying that the Regensburg plant, which he opened in 1937, was originally intended for the production of cars as well as aircraft. Following an unsuccessful attempt to obtain an interview in 1973 I was asked to put my questions to him in writing. Unfortunately there was no response and clarification of this important point is not going to be easy now.

The sad news of his death reached me at Würzburg over the radio at the home of Tiger specialists Bernd and Klaus Eigenfeld while we were having discussions on the work of Fritz Fend as a designer and engineer. This was appropriate since, although Mr Fend has consistently maintained there was no direct connection between the aircraft and the cars of the same name, strong indirect links nevertheless do exist.

The early enthusiasm for aviation of Professor Willy' as he was sometimes referred to in Germany, was aroused by the Zeppelins, but the first practical application of his work as a designer and builder was with gliders, to which until 1922 Germany was restricted by the Versailles treaty.

The need to concentrate on ultra-lightweight construction in the field of motorless flight was to caracterize his subsequent work and I believe we can trace a line from that through to our cabin scooters, which are lighter than comparable vehicles, via the Professor's influence on the work of Fritz Fend which stemmed from their wartime association.

His early experience with gliders also explains why Messerschmitt wasted little time with biplanes when aircraft designers almost everywhere seemed preoccupied with them. In recent times he was asked to sum up his design philosophy in one word and replied "weglassen". That cannot be succinctly translated in English but the meaning is 'to dispense with all non-essential parts'.

Messerschmitt, like Fend, was almost fanatical in his

efforts at reducing aerodynamic drag. Again we can trace a link with our 'schmitts as this, of course, is the explanation of the tandem seating and narrow frontal area. In comparison some contemporary miniature cars were about as aerodynamically efficient as a barn door.

So much has been written about the individual designs of Professor Messerschmitt that it would be superfluous to attempt to add anything to the dozens of first rate books now available. Last summer, though, I was reminded of what was perhaps his most bold and advanced concept of all, the Me 262 jet, through our good friend Heinz Schreiber of Tiger and Nürburgring fame. Acting in his professional capacity as a local government legal advisor in Bavaria, Heinz was called out to the small town of Weicht as a member of a commission appointed to inspect an example of the famous fighter which had lain buried and undiscovered since 1944. While the other commission members were busy making out their reports Heinz quietly 'whipped' a piece off the old Me. which stands on my desk as I write this!

What kind of man was the Professor? Well, great men are not noted for being easy to work with and Messerschmitt was no exception. He was private rather than a public figure who was more at home with his drawing board than when attending business junketings. Although always dignified, on many photographs he gave the impression of being preoccupied and had the appearance of the 'odd man out' when appearing in company with others. No one however should infer from this that he was the typical 'absent minded professor'. Well before the end of his life Messerschmitt had succeeded in swallowing up all his old rivals with the sole exception of the family controlled Dornier firm and emerged on top as a very large shareholder and Life President of the Messerschmitt- Bölkow-Blohm conglomerate. No amateur business man could have achieved that. He never got on with his arch-rival and contemporary Ernst Heinkel and Alexander Lippisch of 'Komet' fame found it impossible to work with him.

Once a team of distinguished Soviet designers visited Messerschmitt and later made unkind comments about him. But success always breeds envy and if you are at the top of your profession, whether you are a Herbert von Karåjan or a Colin Chapman, there will be those who are jealous of your achievements and ready to snipe at you.

In 38 years of reading articles and books about Messerschmitt I cannot recall any instances of him commenting on the work of other designers - with one exception. He may have held on the policy of saying nothing rather than be disparaging. All the more significant then that the sole exception concerns our own Fritz Fend, whose design work has been the subject of justifiably unstinted praise by the Professor.

Not only was Messerschmitt ahead of his time as a designer, when he could he saw to it that excellent conditions were provided for those working for him at ordinary level. Indeed, when in the 1930s a delegation of his British competitors were shown round the plant they were astonished to find such facilities as workers' clinics, swimming baths and canteens already in operation. An echo of this enlightened policy can still be seen in the design of the former administration building in Regensburg's Prüfeninger Strasse, now used by the city as a school for apprentices. It is not easy to believe that these light and airy buildings date from 1937.

Bringing the story up to date, not everyone realises that the modern innovation whereby workers can choose their own hours, known in Britain as 'Flextime", was originated by Messerschmitt-Bölkow-Blohm at the big Ottobrunn plant near Munich.

But Willy Messerschmitt will be remembered longer than any other single individual for his influence on the evolution of aircraft design. We can read his epitaph whenever we look into in the sky and see a modern jet.

With the contribution of the individual designer increasingly being submerged within the 'team' concept, we shall not see his like again.

Tiger

der Vollblut-Sportwagen
mit dem Preis eines Kleinwagens
und der Leistung eines Tourenwagens

ca. **3 400** DM a. W.

je nach Ausführung

TECHNISCHE DATEN DES „TIGER"

Type: „Tiger", Vierrad-Sportwagen.

Hersteller: Fahrzeug- und Maschinenbau GmbH. Regensburg, Lilienthalstraße.

Maße: Länge 2860 mm, Breite 1270 mm, Höhe 1150 mm, Bodenfreiheit 150 mm, Spurweite vorn 1110 mm, Spurweite hinten 1040 mm, Radstand 1885 mm, Eigengewicht ca. 300 kg.

Rahmen: Verwindungsfreier Stahlrohrrahmen mit vollkommen geschlossener Bodenwanne.

Lenkung: Sehr direkte, verschleißfeste Achsschenkellenkung mit Spurstangen in Kunststofflagern. Moderner Lenker mit Signalknopf.

Federung: Einzelradaufhängung mit Progressivfederung.

Bremsen: Großbemessene ölhydraulische Bremsen 180 mm ⌀.

Karosserie: Ganzstahlkarosserie mit wetterfester, gebrannter Kunstharzlackierung.

Innenraum: Hintereinanderliegende schaumgummigepolsterte Sitze. Vorn Bereitschaftssitz, wie er speziell für das Weltrekordfahrzeug entwickelt worden war. Rückwärtige Sitzbank über die ganze Breite des Fahrzeuges, geteilt, hochklappbar. Hinterer Sitzraum Platz für 1 Erwachsenen und 1 Kind.

Bereifung: 4,40 × 10.

Scheinwerfer: 130 mm ⌀ in modernster Ausführung mit asymetrischem Licht.

Leistungsgewicht:

mit 19,5-PS-Motor	mit 24,5-PS-Motor
15,4 kg/PS	12,3 kg/PS

Geschwindigkeit:

Spitze ca. 130 km/h	Spitze ca. 140 km/h

Beschleunigung: Von 0—80 km/h unter 14 Sekunden.

Änderungen vorbehalten!

Interessenten fordern kostenlosen Spezialprospekt „Tiger" F 1 an.

FAHRZEUG- UND MASCHINENBAU GMBH. REGENSBURG

SMALL FRY

You may laugh, but microcar owners take their cars very seriously. Team *C&S* – Chapman, Herrmann, McCarthy, Walsh – met the bubble boys and girls to find out why.
Pictures: John Colley

"...And you must remember", said the concerned voice on the other end of the telephone, "that microcar owners delight in the simplicity and quirkiness of their cars – they're not speed-crazy".

The microcar world was, perhaps unsurprisingly, cautious about the sudden demand from CLASSIC AND SPORTSCAR for as many different types of minimal machine as could be mustered for our big day out. This uncertainty cannot have been helped by our title alone; 'CLASSIC AND SPORTSCAR CLASSIC AND SPORTSCAR – it's all vintage racing and Ferrari Dinos, isn't it?' would be the totally forgiveable thought that might dash through many a Bond Minicar/Frisky Family Three/Messerschmitt KR175 owner's mind.

Nevertheless, with the invaluable help of the Register of Unusual Microcars' Jean Hammond, the energetic Mike Wilsdon and the efforts of Midlands microcar prime mover Kelvin Luty, we began to assemble an array of willing but – we detected – not entirely convinced enthusiasts. They all brought their toys along to Bruntingthorpe Proving Ground so that we could get an insight into just what it is that makes microcar enthusiasm such an obsessive – albeit rather exclusive – arm of motoring.

You can read about what the cars were like 'in action' on the following pages. Trying to identify the

We guarantee you can't buy smaller! Left: at the rear is the Kerry Capitano delivery van, while at the front – in clockwise order from top left – the Nobel 200, BMW Isetta 300, Trojan 601, Bond Minicar Mk G, Messerschmitt KR175, Peel Trident, Scootacar Mk 1, Brütsch Mopetta and Peel P50

typical owner would take rather more explaining, for common traits are not immediately obvious. The cars we sampled were as varied as their owners, but were united in their intention of providing transport for the minimum outlay, not only in terms of fuel and other 'provisions' used but also road space, raw materials and original purchase price. To anyone who has just shelled out hundreds of thousands on a Ford GT40 or Ferrari GTO, purchase price will be anathema, hotly contended before ownership, enthusiastically justified afterwards. Microcars have their own scale of pricing enigmas. You could pay fifty quid for a three-wheeler that can only just accommodate three rubberised midgets and can stomp along at a faster-than-it-seems 40mph in the UK... And then you could dispose of forty times that sum in Germany and end up with a low-line perambulator-style single-seat tiddler that struggles with road camber and could drive over your foot without dulling your shoe-shine.

Money, therefore, is not an asset that has to be gained before taking up micros can be contemplated. Two or three of our feature cars are used daily while many others on the roads of Great Britain belong to one-car families. Collections of microcars are astonishingly common, relatively speaking. After all, you hardly need to have National Car Parks shares to accommodate one or two in the front garden, and indoor rebuilds – often in the bathroom or loft – no longer turn hairs in the Isetta Owners Club.

Cable and bobbin steering and a top speed of 50mph when you could probably get a decent Jaguar XJ6 for the same cash? Bond Minicar owner Andy Price has no qualms:

"It's a social thing really", he told us. "Microcar owners will travel all over the country to rallies and

Top left: the prototype gull-wing Frisky. Top right: special zebra-skin-trimmed Goggomobil T300. Left: unusual Trojan van. Below left: get your kicks with a Messerschmitt. Below right: watery Bond plus equipe. Bottom: Suzuki's 1983 CV-1, one of 50 'samples' – the bubble of the future?

drive in convoy to the continent to take part in European meetings. There's a good feeling among owners and enthusiasts because it's not money or class that counts. You do get a few people who join the clubs just to get spares, but they're the odd ones out".

If it is the improvisation aspect of micro-cars that makes them so endearing today, then it is a strangely inverted appeal in relation to the cars' initial *raison d'être*. In the late thirties, the developed world was enjoying an economic boom after years of post-First World War and Depression austerity. No-one wanted small cars; the trend was to go ever more sophisticated and up-market – and that meant that the cyclecars (in Britain) and voiturettes (in France) rapidly fell from favour to the point where, in 1939 and under gathering war clouds, cars of below-Austin Seven size had all but disappeared.

Following those six long years, industrial empires were in smouldering tatters, *everything* was in short supply except buying power – and car design had remained motionless for half a decade.

First indications that a new sort of economy car was on the way came from France. As a huge number of private cars had been requisitioned by the German forces while they were in occupation, several French stalwarts had built their own electric- and sometimes pedal-powered small runabouts to provide unobtrusive mobility. These had been popular in an underground sort of way but when peace was finally restored, they found a receptive market among the car-starved *Français*, and it was not long before *voiturettes* like the Rovin, Mochet and Julien were selling in quite respectable numbers.

And so the ISO Isetta and the Heinkel immediately earned the nickname of 'bubble cars'

Over in the newly-Westernised Germany, factories that had spent years churning out weaponry were, with help from the Allies, turned over to peacetime production. Three of the bigger aircraft makers, Dornier, Heinkel and Messerschmitt, latched onto the idea of mass-market small car production for their war-torn nation. Dornier's Zundapp Janus eventually came to market in 1956, a weird little four-wheeler with back-to-back seating, but Messerschmitt was much quicker to produce its own car in 1953. An adaptation of a Fritz Rosenheim design for an invalid car aimed at disabled ex-servicemen, the three-wheeled KR175 powered by a 175cc Sachs engine certainly reflected its aero-heritage with its tandem seating beneath a completely glazed, hinged canopy *à la* Me109 fighter plane.

Heinkel's solution had a more normal seating configuration but access was through a single door positioned across the front of the car to which the hinged steering rack was attached. The engine was at the rear, a 174cc unit being of Heinkel's own design and build, as was a single wheel. However, it was the body's shape that caused the most interest, a hunched-up pumice stone outline with huge windows and a neatly tapered behind concealing the engine. Over in Italy ISO's Isetta had also just been revealed and it was of virtually identical concept and style although the engine was considerably larger at 245cc. This and the Heinkel immediately earned the nickname of 'bubble-cars' because of their outlandish looks, an appellation that has lasted to today and has been applied to any car subsequently with three wheels and any trace of curves to its contours.

In England there was the Bond Minicar. This was a true essay in utility with no suspension at the back, tiny seats, little Villers engine that swung around with the steered front wheel and Nissen hut aluminium body. This crude confection was cheap to make and surprisingly easy to sell, such was the pent-up demand for new cars. It inspired lots of imitators too, but the competition like the also-ran Gordon and the dire AC Petite lacked character, tended to be more expensive and were half-heartedly built and sold by companies keen to cash in on Bond's success.

By the mid-fifties, however, and ten years after the Second World War's last shots had been fired, the little fellas were facing increasing competition from the mainstream car makers, whose post-war small car projects were bearing fruit. *Real* small cars could be bought again and the need for austerity had given way to a desire for attractive consumer goods that identified with the new age. Most of the smaller micro-car makers disappeared, Heinkel and ISO stopped producing their machines in-house (but guaranteed themselves a continuing return on their investments by selling production rights all over the place) and even the scrawny Bond Minicar took on full-width styling and pretended to be an Austin with a wheel missing.

The Arab-Israeli war erupted in 1956 and the ensuing Suez crisis ransacked the organised world of oil trading. Fuel supply was threatened and economic disaster loomed.

And suddenly the micro-car was in demand again. Licensees hurried to put the various Isettas, Heinkels and Fuldamobils into production, two or three also-rans gained a new lease of life and there was a spate of newcomers. Some, like the Vespa 400 and Goggomobil and Powerdrive, made serious attempts at real car styling, while the advent of glass-fibre made it possible to produce rival 'bubbles' relatively cheaply; the Scootacar, Peel and Unicar were inexpensive to develop and thus viable even at low volumes.

This time round, the marketing men got together with the economists and the new wave of petrol misers were given bright colours, plaid upholstery, sunroofs and token chromework to give them what today is known as 'showroom appeal'.

The repercussions from the troubles in the Middle East were not, however, as harsh as had been expected, and the revival in the micro-car's fortunes was to be limited. In 1959 the new small car sensation didn't have three wheels, motorcycle engine, glassfibre bodywork or perspex windows; the Mini had everything a small car should have plus the assurance of having been developed by a 'proper' car manufacturer. The front-wheel drive and transverse engine gave safe handling, sporty handling and bags and bags of room – qualities that hardly any micros could boast – and best of all, it was very cheap. Overnight, the improvised economy cars and bizarre shapes were obsolete, their sole and rather weakly-grasped trump card being fuel thrift and even that was a rationale that is now firmly consigned to the past. One by one, the micro-cars dropped away, with the Trojan-built and promoted Heinkel making it to 1965, the venerable Bond Minicar limping on for a year more. The last Fuldamobil left a Greek assembly plant in 1970. The Fiat 500 and 600 quashed the few remaining Italian tiddlers, the French ignored them from behind the wheels of their 2CVs.

Not a lot has happened since then. A few loonies have tried to divert the small car buying public's attention to minimal motoring, wasting their efforts, and sometimes lots of money.

Almost unbelievably, however, the French have relented and there are now some 20 companies engaged in building micro-cars. This fascinating swing has been happening over the last decade and it's spurred on by very advantageous legislation that allows for a 49cc 'car' to be used by a 14-year-old without road tax, licence or insurance. The cars, the Teilhols, Ligiers, Belliers and Erads, use simple chassis, industrial engines of all types and cheap glassfibre bodies – and they are virtually indistinguishable from one another in the main, as the regional caravan manufacturers, central heating purveyors and bus producers who field them employ, how shall we say, 'straight line technology'. When you are designing a car that can be no longer than two metres, how much styling can you create with 200 centimetres of dead straight lines? The French city dwellers, old age pensioners and road-

eager adolescents aren't bothered – they buy these new micro-cars to conserve resources and get about.

The English attitude, however, is not that of our Continental friends. After all, it's power and prestige in Thatcher's Britain, pound-flaunting rather than penny-pinching. Parking meters in London and other big cities destroy any convenience grounds on which the British breed of micro-car could have righteously stood and the decline of the self-contained village community diminishes the primitive attraction of the old Bonds and Scootacars.

Come a crisis, they'll be back though … and in the meantime, if you're really keen to economise, can we recommend the following…?

MESSERSCHMITT KR175

Engine: Fichtel & Sachs 175cc, two-stroke twin-cylinder. **Seating:** two in tandem. **Numbers built:** 2000. **Years built:** 1953-55. **Maximum speed:** 55mph. **Fuel economy:** 80mpg. **Price when new:** £275.

The aircraft manufacturer from West Germany began marketing this Fend design in the UK in 1955. Winford Jones owns one of the first imported to England, and proved his loyalty to the marque by driving all the way from Bristol in the pouring rain, and most of the distance without a clutch. His KR175 is the crudest version of this wonderful family of micro-cars. Unlike the later KR200 and the coveted Tiger, the KR175 has virtually no front suspension and all road bumps are literally transmitted to the handlebar controls. In contrast to the later models, the throttle and clutch are motorcycle-type, and the single brake is foot-operated. The engine is mounted solidly to the chassis so there is plenty of in-car entertainment. In retrospect, earplugs are absolutely essential as are soft-soled shoes, to keep the vibrations out!

In fact the only suspension is in the seats. Although the ventilation at the sides of the cabin is good, even on the shortest of journeys in the summer, the occupants soon start to fry, particularly on top of their heads. The dashboard has a clock and speedo, while the handlebar steering is amazingly direct, great fun and an absolute joy when parking. Thankfully the KR175 is fairly light, as no reverse gear is provided, but four forward gears help produce a rather hairy 55mph.

The most distinctive feature of the KR175 is its styling, with futuristic faired-in front mudguards,

and narrower track than the later models. Star of many feature films and TV commercials, the attraction of the Messerschmitt is obvious in these bland, rational eighties.

KERRY CAPITANO

Engine: Minnerelli 49cc, two-stroke single-cylinder. **Seating:** mono plus load. **Numbers built:** NA. **Years built:** 1965. **Maximum speed:** 40mph. **Fuel economy:** 100mpg. **Price when new:** £125.

Microtrucks and Cargoscooters are a familiar sight in most European countries, but for some strange reason they were socially unacceptable in Britain. Back in the thirties such marques as Scott, Royal Enfield, Reliant and Croft were maginally successful producing light commercials which combined motor cycle and delivery van. But during the fifties various attempts were made to market such vehicles like the Pashley Pelican (which united Royal Enfield motorbike and two wheels behind), as well as Italian moped-based trikes like the Empolini and the Capitano featured here. Both were often seen buzzing round the back streets of Milan or Rome with a full load of cakes or flowers, but in London's Covent Garden there were few takers.

The immensely useful Capitano was marketed in Britain by Kerry's of Stratford, but few were sold in England. Mike Wilsdon saved this example from obscurity for his new museum at Kew Gardens, and explained its attraction. "It's just as logical as a 'bubblecar', so I thought why not befriend something that is even more practical."

Unlike its rival, the Empolini which has shaft-drive, the Capitano is driven by a 9ft continuous chain. With a maximum 5cwt load, and complete frontal weather protection fairing, the Capitano is happy at 15mph. It has a twist gearchange much like a scooter, but the foot brake is mounted on the right, 'continental' style. Handling is stable if treated with respect, but care must be taken when cornering as the rider can forget he has a load behind and attempts to lean the frame, motorcycle-style, have disastrous results. However its turning circle is amazing, being able to turn in its own length.

As an intereting side bar, Mike Wilsdon discovered that early production of Kerry machines involved the East London Rubber company, and Abingdon King Dick Ltd. Obviously transportation for the single man!

PEEL TRIDENT

Engine: Vespa 100cc, two-stroke, single-cylinder. **Seating:** intimate two-seater. **Numbers built:** 75. **Years built:** 1962-65. **Maximum speed:** 50mph. **Fuel economy:** 90-100mpg. **Price when new:** £189 19s.

'Almost cheaper than walking' was the advertising slogan for the weird and wonderful Peel Trident. No doubt the cost of resoling a pair of brogues after a 100 mile hike was far more than a gallon of petrol in 1963. Originally conceived as a shopping car rather than a practical mode of transport, the Trident was built on the Isle of Man. An essential extra was the shopping basket that fitted into the passenger seat, hugely improving roadholding for anti-social drivers. However, its present owner Andrew Carter complained that on continental trips, the only drawback of the Peel is the lack of room for cheap wine.

The rorty Vespa 100cc engine (as opposed to the 49cc economy model), drives the single rear wheel, and provides fairly brisk acceleration at traffic lights, but flats out at high revs around 45mph. Also, the sharp brakes tend to be a little alarming due to the Peel's light weight, but the handbrake is more advisable for straight line stops.

Designed to conform with moped laws, the Peel has no speedo, brake lights or indicators and Andrew continues to have problems convincing MoT testers the car is legal. The single-spoke steering wheel would be ideal for a Citroën DS pedal car, while brake, clutch and accelerator are all floor-mounted, and gearchange is in-line like a motorbike.

Obvious disadvantages are sauna interior temperatures in the summer, and rather alarming 'tram lining' at speed due to the tiny size of the front wheels. Night driving can also present problems because lighting progressively dims on tickover at junctions. Stalling at traffic lights can also be embarrassing because to restart you have to climb out, plug in the starter pedal, kick start, and by the time you're back inside the lights have changed again.

But all the drawbacks of Trident travel are made up for by the amusement value of the 'clockwork orange'. Andrew is not proud, and enjoys the joke as much as anybody, 'I never attempt to defend it, and still laugh every time I drive it.'

SCOOTACAR MK I

Engine: Villiers 9E 197cc, two-stroke single-cylinder. **Seating:** driver in front plus one or two behind. **Numbers built:** 600. **Years built:** 1957-

61 (Mk I). **Maximum speed:** 55mph. **Fuel economy:** 60mpg. **Price when new:** around £350.

The Scootacar, like many another micro-car, had its origins in the corner of a workshop dedicated to something quite different. The Hunslet Engine Company were old-established locomotive builders in Leeds who diversified into bubble cars in 1957.

Malcolm Thomas's Scootacar is a Mk I, built in 1960. Its two-stroke Villiers motor is mounted on a very heavy steel chassis and surrounded by a surprisingly advanced glassfibre moulding. The combination makes it much more stable than it looks from outside.

Malcolm acquired the car in a totally derelict form for a mere £12. He restored it in 1977-78, and the cream interior and exterior is still immaculate. At one time Malcolm had 41 *different* micro-cars (plus some duplicates) but the Scootacar has made it to the final 12 which he still owns and he has driven it to Dutch and German rallies with two people plus camping equipment aboard – and says that a few sleeping bags make it both quieter and more comfortable.

In its Mk I form the Scootacar is probably one of the less comfortable micros, especially for the passenger who sits astride the downstroke of a T-shaped seat; two child-sized passengers could sit side-by-side on the crossbar of the T. The driver sits at the bottom of the T with a rudimentary back support, adjustable fore and aft. Steering is by handlebar and the gearchange is motorcycle-style. It cruises at 45 to 50mph and will keep that speed up all day. But how long can Malcolm stand it? "Once you get used to it, it's really comfortable," he says. "It corners quite fast, when you get wise to it – and driving it is fun."

BRÜTSCH MOPETTA

Engine: Ilo 50cc, two-stroke single-cylinder. **Seating:** solo. **Numbers built:** not known. **Year built:** 1956. **Maximum speed:** 32mph. **Fuel economy:** 100mpg. **Price when new:** unknown.

Looking like a runaway bumper car, the Mopetta was easily the rarest of the micro-cars present at our test day. Conceived in Germany by Egon Brütsch, who had a reputation for never completing a design

before he went on to something else, the Brütsch was designed as an amphibious vehicle. The trike utilised the driving motion of one wheel to propel it through water. Unfortunately the Brütsch only succeeded in sinking!

However a range of prototypes was constructed with configurations of three or four wheels, solo or two-seaters and various power units including Ford Taunus. German authorities refused to register the Brütsch for road use, so Egon sold the remaining models to England. They were unsuccessfully marketed by Raymond Way Motors of Kilburn.

The Brütsch has a pull-start just like a lawn mower, and once aboard, the fearsome 50cc offset engine is operated by handlebar controls with three gears, and a floor-mounted brake pedal. The fashionable see-through polyethylene hood is erected when sitting in the cockpit, and has to be folded down to get out. A novel feature of the chassis is the lever arm suspension which has rubber band-style damping, providing a very soft, springy ride, and alarming handling.

Often stated as the smallest car ever built, the Brütsch is just 5ft 7in long, and Malcolm Goldsworthy summed it up perfectly as "the most useless car and boat at the same time", but if your lifestyle requires the most exclusive micro-car built it has to be the Brütsch or... possibly... a Volugrafo Bimbo...

BOND MINICAR MK G TOURER

Engine: Villiers 249cc, two-stroke twin-cylinder. **Seating:** two abreast. **Numbers built:** 90 approx. **Years built:** 1964-66. **Maximum speed:** 55mph. **Fuel economy:** 45mpg. **Price when new:** £408 19s 10d.

Laurie Bond has been described as a 'weight-saving maniac'; his first Bond Minicar, offered to a car-desperate public in 1948, was certainly light on rear suspension (it had none) and the cable and bobbin that made up the steering wasn't on the heavy side either.

Even by the sixties, however, 'Bonding' was a pastime that relied on minimalism, and Andy Price's 1964 Mk G tourer shown here is no less an economy machine than its more rotund adversaries. The big car styling hides a surprising amount of nothing; beneath that full-width 'pontoon' is enough room for the engine and front wheel to swivel through 180 degrees, giving the Minicar a turning circle of precisely twice its length. The rest of the body – not

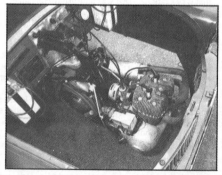

including the bendy glassfibre lids to engine and boot – is in aluminium.

Still rather dizzy from spinning round and round and round (the turning circle really is incredible) we set off for a drive – and it was rather impressive, actually. The 197cc motor pulls capably and even at a somewhat worrying 50mph, civilised conversation is possible. With one wheel at the front, a good deal of motorcycle-type leaning is advisable in corners, to avoid sliding to a halt on one side, but fortunately, there's lots of room on the upright bench-seat to do so. The ride over normal, slightly uneven road surfaces was also well up to Austin A30 standards.

The name's Bond, but I was stirred not shaken.

TROJAN TYPE 601

Engine: Heinkel 198cc, four-stroke, single-cylinder. **Seating:** two-plus-two children. **Numbers built:** 6000 approx. **Years built:** 1957-65. **Maximum speed:** 56mph. **Fuel economy:** 65-110mpg. **Price when new:** £380.

The Trojan, *née* Heinkel, was one of the more successful micro-cars, being made in Germany from 1957-59, in Dundalk (Ireland) from 1959-62, and in Croydon from 1962-65. It was, along with the Isetta, one of the true bubble cars – a glance at the ovular shape makes the point. And, like the Isetta, there's that forward-opening door…

The Trojan has a 'proper' layout, with two wheels at the front and a singleton at the rear, *á la* Morgan: in this case, triangularity leads to stability. It also leads to a spacious cockpit, for two at least: the '+2' bit in the rear is really for children under the age of one… Steering is by a proper wheel, brakes are properly hydraulic, and the only motorbike connection is the gear lever with a sequential gate – all changes are fore-and-aft. No H-patterns here.

Kelvin Luty, the owner of the Trojan, has been a micro-car fan for a long, long time – he owned a Berkeley while he was still at school. However, it took his wife to convince him to buy a Trojan – as a total outsider to the scene, she loved the space and the shape. So Kelvin bought one, then another, and another, until he had 12! He's managed to reduce the numbers now, though, more than somewhat. His longest journey was from Leicester to London – then back up to Edinburgh, emulating a trip Trojan made in 1962 to prove the device. He completed the London-Edinburgh section in 12hr 20min, and recorded 73.5mpg in the process. That was with his wife, dog and camping gear: on a one-up trip to Bristol he's managed 100.6mpg.

To drive, the Trojan is a total surprise. All the controls are big-car, gearchange excepted. Performance is amazing for such a tiny engine – on a level straight you'll see 40mph fairly rapidly, much quicker than you'd expect in fact, and it'll cruise at 50mph as happy as you like. The steering was a mite sticky for my taste, but at least that means that the car doesn't dart around the place as you might expect. The best part of the whole thing, though, was the gearchange: talk about electric switches! Click, one, click, two, click, three – superb.

NOBEL 200

Engine: Sachs 191cc, two-stroke single-cylinder. **Seating:** two-plus-two. **Numbers built:** around 250. **Years built:** 1959-61. **Maximum speed:** 55mph. **Fuel economy:** 45-50mpg. **Price when new:** £260.

The Nobel 200 is basically a Belfast-built Fuldamobil – although the Sachs engine is the same one used in the Messerschmitt. Indeed, the only way to get hold of engine parts in Britain is through the Messerschmitt club – but you can't join the club unless you own a Messerschmitt, and there aren't any spare Messerschmitts to be had…

Built, albeit briefly, by shipbuilders Harland and Wolfe for industrialist York Nobel (and abortively by Lea Francis), the Nobel's most striking feature is its two-tone paintwork: cream above the zigzag trim line, turquoise below. The body is glassfibre, the chassis is steel and the floor is wooden (beware woodworm).

Ray and Jenny Dilks only got their Nobel on the road last year – they bought it, in semi-restored state, from Malcolm Thomas. There's room for two adults in front and two very short people behind (headroom is the limitation). The doors are at the sides and the gearlever is in the middle; it looks conventional but in fact it operates a motorcycle gearchange and comes complete with neutral lever.

BMW ISETTA 300

Engine: BMW 295cc, four-stroke single-cylinder. **Seating:** two-and-a-half abreast. **Numbers built:** 30,000 in Britain. **Years built:** 1958-63. **Maximum speed:** 55mph. **Fuel economy:** 50-60mpg. **Price when new:** £306.

Originally produced – for a mere three years – by ISO in Italy, the great majority of Isettas were made by BMW in Germany; production was also carried out in France (as the Velam), Belgium, South America and Brighton, England, in a factory that was only reachable by rail.

Although its looks are distinctly bubbly, the Isetta is more car-like than you'd expect, not least because it has two wheels at the back. They're only 18in apart, but that's three inches too far apart to qualify them as a single wheel in Britain, so Isetta owner Helen Ayriss has to fork out a full £100 a year for a tax disc instaed of £40 for a three-wheeler. For this reason, although almost all of the Isettas sold in Europe were four-wheelers, most of those sold in Britain were three-wheelers. It was a tax-saving 'dodge' that may, in the end, have saved on fuel and tyre wear, because there's no differential between the rear wheels, so on a bend the outside wheel is simply dragged round.

On Helen's right-hand drive, Brighton-built car, the gearchange is conventional H-plan. On left-hand drive models, however, the whole gearbox and linkage were simply mounted upside down, so the gearchange is laterally reversed.

There's plenty of room for two adults on the exceedingly springy front bench seat, plus one small child in between. The combination of 295cc four-stroke engine and hydraulic brakes make starting and stopping a doddle. It's taking right-angled bends at 15mph that's a bit tiresome. "I get so used to it," says Helen, "I start driving my Renault estate the same way – and then my kids shout at me…"

PEEL P50

Engine: DKW 29cc, two-stroke single-cylinder. **Seating:** single. **Numbers built:** 45 approx. **Years built:** 1962-63. **Maximum speed:** 40mph. **Fuel consumption:** over 100mpg. **Price when new:** £199 10s.

There *are* taller people than Malcolm Goldsworthy. But not much taller. Yet not only is Malcolm micro-car-mad, but there is enough room within this modern-day sedan chair that is the chair-in-box-on-wheels that is the Peel P50 to accommodate his six-foot-and-the-rest stature.

Often billed as the world's smallest car (not true – Malcolm's Brütsch Mopetta scuttles off with that acolade) the Peel P50 was the first car to emanate from its maker's Isle of Man works, and was designed to fill up the gaps between city-centre parked cars. Principal features include a large, French-window-style handle at the rear to enable nimble commuters to drag the car into place, a thinly-padded tubular seat, a REAL steering wheel (there's no play in its 90 degree turning capability) and a 49cc DKW moped power-pack wittering away beneath the driver's posterior.

Stepping through the single door is simple and the seat is comfortable despite its summer-house appearance. Grasping the wheel and working your way through the motorcycle-type gears presents no problem. Indeed, the Peel's performance is most affected by outside factors like wind, any sort of incline or uncontrollable mirth. Thirty-five miles-per-hour is a possibility but fuel is sipped at the rate of some one gallon for every 90 miles travelled.

3-WHEEL CARS

Some answers to questions we forgot to ask

BY RANDOLPH BECKMAN

ILLUSTRATION BY BILL DOBSON

CRAZY JOHN DRIVES a 3-wheel car. He likes it because it is practical, not because of all the attention it gets—which is almost more than he can stand. On freeways he finds himself boxed in on all sides by gawkers. In busy parking lots it draws a crowd of curious touchers. And at *every* traffic light, the driver of the car stopped next to him will roll down his window, look down, and ask one of just two questions: "What is it?" and "Why does it have only three wheels?"

Only the driver of another exotic car can comprehend the inevitability of these questions. Crazy John could answer them from the depths of sleep. Unless the asker happens to be a very sexy person of the opposite gender, the conversation goes precisely like this: Next car: "What is it?" John (casually checking out the asker): "I built it myself." Next car: "Why three wheels?" John (long pause, thoughtful expression): "Why do *you* need *four*?" The word "four" is always carefully timed to coincide with the green light, and away drives John, leaving the counter-question echoing in the other driver's ears.

Why indeed? Three points determine a plane. A fourth support is not only redundant, but it creates a "statically indeterminate structure." In other words, it is difficult to predict the actual load distribution among four suspensions. If you have four passengers, you arrange them in a rectangle, put a box around them, and put a wheel at each corner. But what if you only have one or two seats? Why add the air drag of the fourth corner and why add the cost and weight of the redundant wheel, tire, brake and suspension? Even ignoring the physics of the situation, state and federal regulations seem to encourage 3-wheel vehicles. They label them "motorcycles," allow reduced license fees and taxes, and require fewer safety and emission-control features that add cost and weight.

Before we take a good look at the *dis*advantages of 3-wheel cars, let's test our prejudices and preconceptions. Before you go any further, make a mental note of your gut responses to these common beliefs:

(A) 3-wheel cars overturn more easily.

(B) 3-wheel cars have unique handling characteristics.

(C) 3-wheel cars have a tendency to oversteer.

3-WHEEL CARS

(D) 3-wheel cars have poorer braking characteristics.

As R&T noted in a story called "Are Three Wheels Enough?" in March 1955: "The question that most people want answered is whether the elimination of one wheel is dangerous." Surprisingly, the answers to all these questions finally came not from the world auto industry, but from a federal research program.

When the U.S. Department of Energy (DOE) was formed, conservation-minded innovators inundated it with far-out ideas, requesting research funds and loan guarantees for their projects. Among these ideas were many versions of the electric 3-wheel commuter car. DOE, not wanting to risk a Golden Fleece Award, said in effect that there must be something wrong with the idea, because not only did Detroit not build 3-wheel automobiles, Japan didn't *either*. Walt Dippold at DOE called the Department of Transportation and asked the National Highway Traffic Safety Administration to look into it. Dr Joe Kanianthra at NHTSA soon discovered that while it was possible that 3-wheel cars were less stable, there wasn't any hard evidence to *prove* it. So NHTSA let out a small contract for someone to run down the facts. Fortunately, the research was done by an experienced car enthusiast. Paul Van Valkenburgh is a former auto writer, race car engineer, builder and driver. When he did the project at Systems Technology, Inc, he tested eight 3-wheelers, four with a single front wheel and four with a single rear, and he compared them to four roughly equivalent 4-wheelers.

With 20/20 hindsight, R&T probably could have come to the same conclusions using its own testing procedures. But given federal funding, a wider range of 3-wheel examples to select from and far more extensive tests, the results are not only more comprehensive, but essentially bulletproof. The final technical research report runs to more than 100 pages of explanation, graphs and detail, but I can translate most of it from sophisticated tech-talk and bureaucratic dullness into plain English.

The first concern of those who insure, if not those who drive, 3-wheelers is the possibility of turning over. A theoretical dissection of the problem turns out to be less important than the simple observation that one test 3-wheeler had better overturn resistance than the best 4-wheeler tested, a Fiat X1/9. Sound ridiculous? It isn't if you look at the numbers. If you get the center of gravity

(cg) low enough and close enough to the 2-wheel end, and have a wide enough track, you can build a 3-wheeler that won't overturn in the most extreme maneuvers on flat pavement. Naturally, if you retain the identical cg location and track width of a given 4-wheeler, three wheels will provide a lower overturn safety margin. But these design parameters are never cast in stone and, in limited production, you can put the cg and track just about anywhere you choose.

To be honest, the 3-wheeler with the best overturn resistance was *sans* body, and therefore had an unrealistically low cg. But another 3-wheeler, fully equipped, was still almost as overturn resistant as the Fiat. Also, I should note that one of the 3-wheelers *could* have been overturned in testing and in fact *was* by its owner. It had a high, rearward cg and narrow track, and at 0.6g it would go up on two wheels, precisely as predicted from static tests. To a skilled driver, this was no problem, and it could be balanced there like a motorcycle as long as you had enough time, space and presence of mind.

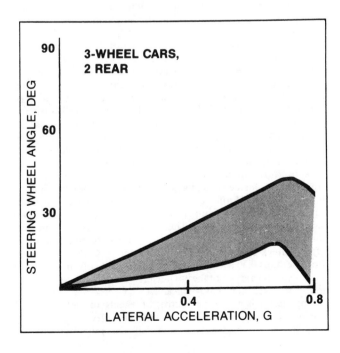

Band widths represent range of handling behavior for a variety of 3- and 4-wheel cars that were tested. Note that 4-wheel and single-rear 3-wheel cars understeer at the limit and that single-front configurations exhibit oversteer.

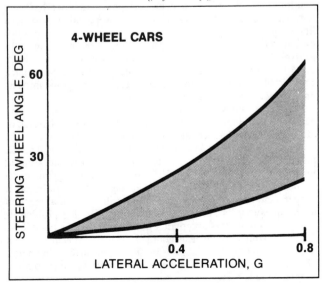

3-WHEEL CARS

But now I come to a philosophical question. Is a vehicle that will overturn during hard cornering "unsafe"? Perhaps it is even more of a *legal* question, as there are some current lawsuits trying to determine if Jeeps and other recreational vehicles are "unsafe" in overturn. If overturn in cornering is unacceptably dangerous, what about tall, narrow commercial vehicles, such as loaded tractor-trailers, which can go into oversteer at about 0.3g and overturn at about 0.6g?

Even if I assume that engineers of 3-wheel automobiles would optimize the overturn design limits, I still need to know how their stability and handling compare. I will ignore the myriad ways of analyzing stability (static, dynamic, transient, steady-state, oscillatory, divergent, convergent) and just consider oversteer/understeer. Most car enthusiasts have an intuitive feel for these terms and respect for the potential dangers of unexpected oversteer at the limit.

As it turns out, there was a strong distinction between 4-wheel cars and single-front-wheel cars—but not single-rear-wheel cars. Although the sample was admittedly small, the inescapable conclusion is that all single-front 3-wheelers will oversteer at their limit of adhesion. Conversely, all single-rear-wheel cars had strong understeer at the limit. And in neither case could the opposite effect be created, in spite of all the chassis tuning.

An accompanying graph shows how all three groups of cars compare in steer angle versus lateral acceleration on the skidpad. Conventional 4-wheelers have a constantly increasing steer angle as speed or g-forces increase. The same is true (to a potentially greater degree) with single-rear 3-wheelers. But the steering on single-front 3-wheelers levels off and then *decreases* with increasing g's, requiring counter-steering to avoid a spin.

If these results were difficult to predict, they are easy to explain after the fact. Oversteer/understeer is a result of many vehicle dynamics factors, such as tire size, type and pressure, suspension characteristics, steering compliance, weight distribution and roll resistance distribution. With all other factors being roughly equal, the end of the car with the greatest weight and greatest roll resistance (springs or anti-roll bar) will have the lower limit of adhesion. Put another way, a nose-heavy car or one with lots of its roll resistance up front will understeer; and vice versa for rear/oversteer. The implications are obvious. With a single front wheel, most of the weight is at the rear, not to mention *all* of the roll resistance.

On some of these 3-wheelers, extremes of state-of-the-art chassis tuning tricks were attempted, with negligible effect. Regardless of large tire and pressure differences, camber changes and changes in weight distribution that were unreasonable from an overturn standpoint, they still oversteered. But again, is oversteer unacceptable? There are a lot of naive folks driving around out there in oversteering production sedans (because of low rear tire pressures) who will never encounter that limit even in an emergency. The *fact* of oversteer is easy to obtain; the implications are more than a little speculative.

So overturn can be avoided, and single-rear 3-wheelers have a comfortable degree of understeer. But what about handling— how do they *feel*? Professional researchers resist being quoted on subjective impressions, but at least here they report a numerical value for handling response. These yaw response times represent the time required for a car to reach a steady cornering condition after a quick steering wheel input. Ordinary 4-wheelers range from perhaps 0.30 seconds for a large car with soft tires, to 0.15 sec for sports cars. All of the 3-wheelers were below 0.20 sec, to as low as 0.10 sec, and that is quick.

The answer is not in the number of wheels, or their location, but in mass, tires and polar moment. The effect of polar moment has been considered for years, but this is the first report I have seen with actual figures. The 3-wheelers had, on the average, about 30 percent less polar moment (normalized for weight) than 4-wheelers, because of centralized masses and less overhang. And the ones with the lowest figures and best tires had the quickest response. Van Valkenburgh says that some of the 3-wheelers had yaw characteristics akin to those of formula cars.

All the rest of the tests showed no measurable difference between 3- and 4-wheelers. The testers subjected the cars to crosswinds, bumps in turns, braking in turns, free steering return, lane changes and off-camber turns, and although there were many vehicle-specific problems (as you would expect with one-off prototypes), the number of wheels was unimportant.

However, all this research, original as it was, had a relatively narrow orientation. There is more to a practical automobile than handling and braking. Although most design factors, such as quality, esthetics, value, reliability, comfort, etc, are unrelated to the number of wheels, there are still a few other important considerations.

It was assumed that a 3-wheeler could be inherently more aerodynamic than a 4-wheeler. As a part of this research, two of the more complete 3-wheelers were coast-down tested for aerodynamic drag. In both cases, the product of the frontal area and drag coefficient ($A \times C_d$) was about 8.0, as compared to 8.4 for a Plymouth Horizon or 8.8 for a Honda sedan or Ford Fiesta. Certainly this is an improvement, but not as good as the 6.5-7.0 for comparable low-drag prototype 4-wheelers. This peculiarity partly results from the necessary increase in frontal area on a wider-track 3-wheeler and partly from the lack of aerodynamic optimization on any of these cars. Obviously, air drag of a 3-wheeler will not automatically be less.

Another theoretical justification for a 3-wheel chassis is that it can be lighter because it requires no torsional resistance as one wheel rises and falls. In fact, however, one builder reported that he had a torsional stiffness problem on one of his 3-wheelers. While the single rear suspension was strong enough to resist the camber forces in hard cornering, the frame was not and it not only flexed in torsion, but allowed disturbing vibrations.

Ride quality can be a problem as well, not in vertical ride rate which is independent of the number of wheels, but in 1-wheel rates. Because the 2-wheel end of the car must resist all roll, roll resistance requires more stiffness than necessary for a soft ride. And, as Van Valkenburgh discovered, roll *damping* suddenly becomes an important consideration when all of it has to be handled at one end.

Finally, all this research ignores the question of impact protection, and rightly so, because it's hard to imagine the number of wheels being a factor. However, the intent of a 3-wheeler is to have reduced mass, and a side effect of three wheels is a "motorcycle" classification, which means that federal impact regulations do not apply. Again, it's a philosophical question. Is a lightweight 3-wheeler a safer motorcycle, or a less safe automobile? NHTSA is currently struggling with the problem. Subjectively speaking, I can't see why a front-engine front-wheel-drive 3-wheeler is inherently any less safe in frontal impact than a comparable 4-wheeler, although crush space or diagonal impacts on a corner with no wheels could be problems.

Now that fwd subcompacts are flourishing, it would seem natural to adapt them to three wheels, either in original design or aftermarket kits. The technical problems involved in producing a practical 3-wheel car do not appear to be overwhelming. And the potential benefits of cost and fuel conservation would seem to make it worthwhile. As Van Valkenburgh succinctly put it, "a properly engineered 3-wheel car can be made as stable as a properly engineered 4-wheel car." But recall your initial reaction to the questions of stability and handling. The *big* problem is *psychological*—market acceptance of a radical change. Even if the 3-wheel layout were twice as good, I wouldn't speculate about its future. One of the most powerful forces on earth is the inertia of an existing idea. But if 3-wheelers ever have a chance to make it, their time is now.

by Rob Maselko and Marsha Wenk, Wharton, NJ

The Messerschmitt KR-200 "Cabin Scooter", once simply an economical form of minimal transportation, is now considered a classic design of "advanced" minicar. The KR-200 is a development of the earlier KR-175, and features a larger engine, wider track, and more ornamentation and interior trim. Our test car is an original, unrestored 1956 DeLuxe UK model, now in the hands of its second owner. The original owner put 50,000 miles on this car before it was put into storage in 1967, which is an amazing number of miles considering the diminutive size of the car, but it is a testimony to the car's usable design and rugged nature.

When initially confronting a Messerschmitt, one is struck by the fact that it is like nothing else on the road (a serious understatement!) You could make a case for it's being a car, a scooter, or a go-kart, but in fact it is all of them and none of them. It is really a unique class of vehicle, with possibly the only other car like it being the British made Scootacar. Surveying the outside of the Schmitt, the tiny size might make you concerned about being involved in a collision with another vehicle, but we will leave that up to the individual contemplating the purchase of a Messerschmitt, and go on to say that it is 112 inches long, 49 inches high, and for the most part, very narrow, except for the proportionately adequate front track.

Entry into the car is accomplished through the right-side-hinged canopy assembly which is opened with a twist of the keyed latch. While it is not very heavy, the canopy is of sufficient weight that it is best if one's feet are firmly planted alongside the vehicle before attempting to open it, particularly if one is not overly muscular. A leather strap serves to keep the canopy from swinging open too far to the right and smashing the dome. With the canopy open, the entire interior and any luggage or passengers contained within, is exposed to the weather.

The driver's seat is mounted on a parallelogram frame which has two positions: a raised position for entry and exit, and a lowered position for driving. Before getting in, you raise the seat, then step in and sit down. Supporting yourself on either side of the canopy, you then lower yourself into the driving position (nearly sitting on the floor). Getting out of the car is a little more challenging, especially if you are wearing a slippery jacket, because you must push back and up in the seat while lifting with your legs and arms, and viola!, the seat returns to the raised position for getting out of the car. This is not a car for the arthritic or handicapped. Like many other aspects of this unique car, it takes some familiarization to get used to, but it eventually becomes an easy and natural way to get in and out of the car.

Once seated, you reach around to the right side of the back seat to twist the fuel tap into the "On" position, pull the manual choke (for cold engines), and turn the ignition key to operate the starter. Again, more explanation is needed here. The ignition switch has two positions. One is the normal "twist and start", but the other requires you to first depress the entire ignition lock tumbler about 3/8ths of an inch, then twist to start the engine up in reverse! You see, the motorscooter derived Sachs 200 drivetrain does not have a proper reverse gear, so instead you start the engine up backwards to enable you to back up the car when engaging first gear. A four stoke engine is incapable of running backwards (except, sort of, when you shut it off and it runs on from being fed a diet of low grade fuel), but a two stroke engine can start up in reverse when

equipped with the proper electrics and separate ignition circuit.

The Sachs fires up quickly with a familiar two-stroke ring-a-ding sound, requiring just a little choke from cold. It warms up in just under a minute, and the improvement in throttle response tells you when it is ready. If you have ever driven a motorcycle, then you will readily grasp the shifting routine. If not, then here is another area which will require some practice before you will feel comfortable enough with the mechanism to venture out into traffic. To use a well-worn road tester's cliche, the shifter itself is well located and falls readily to hand. (The shift knob is "adult" sized, as opposed to the tiny, slender knobs found in other bubblecars.) It only moves fore and aft, not side to side, and has three positions: center, backward, and forward, with a return spring pushing it back to the center.

The clutch, brake, and accelerator pedals are floor mounted and operate in a conventional way. There is plenty of room around the pedals for this tester's large 10-1/2 shoes, and there is even a thoughtful "dead pedal" to the left of the clutch for when the left foot is not in use. You engage first gear by depressing the clutch, then firmly pulling the gear lever all the way back. You not only will feel the gear engage, but you will often hear a light "crunch", which is perfectly normal for this type of transmission. The gear lever will spring back to the center position, while the gear remains engaged in first. Using plenty of throttle, you gently slip the clutch as it engages, and you are underway. Second gear is engaged by declutching as normal, and then firmly pushing the shift lever all the way forward, allowing it to then return to the center, and so on up through the gears. When coming to a stop, you can either declutch and stop, then pump the shift lever backwards to return to first gear, one gear at a time (4 to 3 to 2 to 1), or begin to downshift through the gears as you approach your stop. I prefer to downshift so as to be ready to resume forward acceleration if suddenly required. When you have reached first gear, simply give a pull to the finger trigger mounted on the shift lever, and the transmission will be dumped into neutral,

just ahead of first gear. This neutral selector is only about 85% reliable, even when properly adjusted, so some caution is advised when, thinking you are in neutral, you begin to let up the clutch pedal.

Once underway, the Schmitt feels quite lively in performance, although the stopwatch tells another story. The factory gave the top speed as 62 MPH, and the owner tells us he has had the car at that speed on a slight downgrade, however, as with most microcars, you could probably clock the performance with a sundial (0-30 in 23.2 sec. 0-40 in 32.4 sec.). In local traffic, the Messerschmitt manages to extract enough acceleration from all of its 10 horsepower to keep up with the normal pace provided you make full use of the shiftlever and the willingness of the Sachs 200 to rev freely. Performance is greatly affected by hills, elevation, humidity, temperature, and whether the car is carrying only a driver, or a driver and passenger. The lack of serious horsepower obviates the need for modern electronic traction control on slippery surfaces, however at the track, it hampers violent acceleration. The car understeers strongly in tight, slow speed turns, but becomes negligible at speed. Our testers found that it was possible to lift the front inside wheel during hard cornering, but it was easily corrected with a flick of the steering bar, and at no time does the car feel unstable. For a three wheeler, it is very tossable into corners, a tribute to Fritz Fend's design, where the driver actually sits in the car's very low center of gravity

Subjectively, what is it like to drive a KR-200? In a word: fun! It gives you the illusion of quick acceleration, the engine makes all the right sounds without being overly intrusive, and the steering is immediate. All our test drivers remarked about the "fun-to-drive" factor. The handlebars (or "steering bar" as is noted in the factory literature) are comfortable to hold, and with a 1:1 steering ratio, the driver does not so much as turn the bar to go around corners as much as think about it. One tester quipped that the steering felt and responded like the

steering bar of a well seasoned American Flyer sled! Ergonomically, most controls are well placed, easy to use, and have a decent tactile feel. The driver's seat back has a nice curve to it that parallels the shape of this tester's spine, offering surprisingly comfortable support. Instrumentation includes an odometer, a speedometer, and as this car is a DeLuxe model, the normally optional spring wound clock. Warning lights are employed for high beams, turn signals, Dynastart charging, and forward/reverse ignition. The dimmer switch is the old fashioned, foot operated button type, located to the left of the dead pedal on the floor. Needless to say with a plexiglass dome, there is 360 degrees visibility; no blind spots at all. Keeping in mind that this was meant to be an ultimate economy car, fit and finish, and quality of materials is of a very high order. Since the test car is an early KR-200, it has genuine wood trim flanking the insides of the canopy; no veneer here! This feature disappeared on later models, along with the chrome headlight rims (later cars used aluminum), as cost saving measures.

The brakes are cable operated on all three wheels, and while they work fine, the feel you get through the pedal confirms their mechanical nature. We noted some fade after extensive use down a large hill (engine braking with a two stroke is virtually nil), but not so severe as to pose a safety risk. By modern standards, the brakes are insufficient. The extremely long stopping distance from 30 MPH (64 feet) indicates one would be best off to anticipate the need for braking. The brakes squealed when a light touch was used, but were quiet under heavier braking. We suspect that the brake cams and front wheel bearings are ready for replacement. The parking brake, operating on the rear wheel, held the car on any hill we could find.

The Messerschmitt uses rubber-in-torsion in place of steel springs, and suspension travel is very limited. Small bumps and road irregularities are absorbed nicely by the rubber, but pot holes and frost heaves send a major jolt through the frame and passengers. We found while testing, that serious bumps can cause a mildly annoying phenomenon: the jolt from a bump can cause the driver's foot to be momentarily lifted from the accelerator pedal, accentuating the effect of the bump. It is important to keep the tire pressures at the factory recommended settings (15 psi front/ 30 psi rear and spare) to get the best results in terms of ride, corning, and braking stability.

As you might expect, luggage space is at a premium. The spare tire and jack are stowed in the engine compartment, under the floor of the "trunk", which is small, and located in the top section of the engine cowl. It will only hold a small briefcase; soft luggage will make the most use of the space. Be sure you put only heat resistant items in the trunk, because the engine makes the trunk quite warm, even in cold weather. The optional luggage rack is recommended.

Passenger space on the other hand, is excellent. The driver seat has nearly a foot of travel, and legroom, headroom, and elbowroom is abundant. The passenger has virtually unlimited legroom because their legs straddle the driver's seat, and it is possible to put a medium sized adult and small child in the rear bench seat. Rear seat headroom is ok, but tall passengers may find the back of their head hitting the dome. The rear seat cushion has an 85-15 split with the smaller portion flipping up to accommodate a piece of luggage on the floor. The entire seat cushion can flip up to carry large items, or as demonstrated by this car's owner, several bags of groceries.

Should you buy a Messerschmitt? It depends on your priorities, and what you expect to get from an automobile. Thirty six years ago, the KR-200 was an effective and economical means of going about. Today, it is a lot of fun to drive where there is minimal traffic and smooth roads. As a microcar, it is one of the best from the point of space efficiency and performance. If having fun as a result of driving means anything to you, and if you don't mind driving a major attention getter, this could be for you.

Our recommendation, of course, is if you find one for sale, BUY IT!

Air filter: Wet filter with induction silencer

Drive sprocket on mainshaft: endless chain, 44 links
12,7x8,2x8,51 roller diameter

Lubrication: Fuel mixture 25:1
(Fuel and two stroke oil SAE 40-50)

Gear Lubrication: approx. 750 cc. gear oil.SAE 80
(change every 10.000 km (6.000
miles) or at least after one year)

Gearshiftmechanism: Hand ratchet via noiseless Teleflex
cable

Exhaust silencer: Baffle

Exhaust pipe: 38 mm inner diameter

Weight of engine including exhaust system:
approx. 35,5 kg (78 lbs)

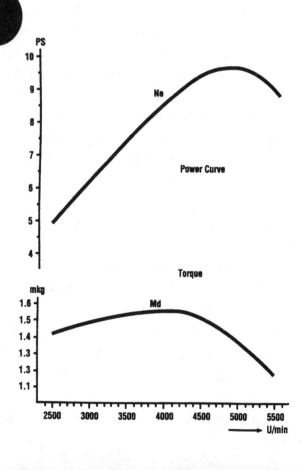

8 • Electrical Equipment: BOSCH-Dyna-starter LA/DAQ 90/12/1700 + 02 LR 2, 12 Volt, 90 watts. Two 35 watt head lamps, 105 mm (4 inches) diameter with main beam pilot light on dashboard. Two rear lights incorporating reflectors and brake lights. Number plate light. Two "blinker" signalling lights disposed either side of the vehicle with control on dashboard. Parking lights*, switchable right or left-hand. Electrically operated windshield wiper. 12 volt battery, 18 Ah.

9 • Final Drive: Cardan intermediate shaft from engine to the rubber mounted independently suspended swinging arm, roller chain 1/2x5/16" in oil bath driving rear wheel.

Lubrication: approx. 650 cc gear oil SAE 90 (change every 10.000 km = 6,000 miles or at least after one year).

10 • Performance: Maximum speed up to 100 km/h = 62 m. p. h. Climbing ability 33% Standard fuel consumption approx.. 3,4 ltr. (= mpg UK or 69 mpg US). Average fuel consumption on the road 4,2 ltr./ l00km**(= 66 mpg UK or 56 mpg US)

11 • Fuel Tank: Accessible from outside the vehicle. Capacity approx. 13 ltr. of which approx. 2,5 ltr. in reserve, sufficient for 40 to 50 km = 25 to 30 miles. Changing to reserve effected by means of the fuel tap. (Tap positions are: upwards, reserve, sideways off, downwards on. Tap is located near the rear seatback

12 • Controls:

Hand controls: Steering bar
Right Hand: gear change, neutral selector (on gearchange), handbrake
Left hand: starter, blinkers, light switch, parking lights.
Windshield wiper switch, horn, fuel tap and heater control.

Foot controls: *Right:* accelerator, brake pedal
Left: Clutch, dip-switch

* For English customers the parking lights ore placed on the wings.

** 1 mile = 1,6 km; 1 gallon (UK) = 4,5 Liter; 1 gallon (us.) = 3,8 Liter.

Technical Specifications

1 • Body: All steel body with weather resistant stove-lacquered finish. Rigid steel frame with fully enclosed floor. Tandem seating with foam rubber upholstery. Front seat is adjustable and can be tipped up: full width rear seat, right half removable. Rear seat accommodates an adult and a child.

2 • Steering: Direct steering with wear resistant track rods in plastic housing. Modern steering bar with horn.

3 • Suspension: Independent suspension; soft rubber torsion springs with hydraulic shock absorbers on all wheels.

4 • Brakes: Internal expanding brakes operating on free cable system. Foot and self-locking hand brake operate on all three wheels.

5 • Tires: 4.40x8" all weather tires. Interchangeable wheels. 4.00x8" Special Equipment for England.

6 • Dimensions:

	KR 200	CL*	KR 201	Sport
Length	9' 4"	9' 4"	9' 4"	9' 4"
Width	4' 2"	4' 2"	4' 2"	4' 2"
Height	4' 1"	4'	4'	3' 2"
Track	3' 7"	3' 7"	3' 7"	3' 7"
Wheel base	6' 10"	6' 10"	6' 10"	6' 10"
Ground clearance	6"	6"	6"	6"
Unladen weight	506 lbs.	510 lbs.	510 lbs.	466 lbs.
Laden weight	946 lbs.	950 lbs.	950 lbs.	906 lbs.

7 • Engine: SACHS 200 L-AZL-R with fan cooling

Engine type:	Air-cooled single-cylinder two stroke
Scavenging principle:	Loop scavenging
Bore:	65 mm
Stroke:	58 mm
Capacity:	191 c. c.
Compression ratio:	6,3 :1, to the total stroke
Output:	97 b h p at 5250 rpm
Transmission:	4-speed-gearbox in engine block
Clutch:	4 discs
Idler gear shift:	Hand lever via bowden cable
Starter:	Dynamo starter 12 volts, 90 to 135 watt, output for lights, with electric reverse gear
Timing advanced:	Forward gear: 4,5-5,5mm before TDC Reverse gear: 3-4mm before TDC
Contact breaker gap:	0,4-0,5 mm
Spark plug:	Bosch M 225 P 11 S for medium stress, spark gap 0,7 mm
Carburetor:	Bing starter carburetor, single valve, 24 mm dia. with choke
Carburetor settings:	Main jet 120, needle jet 1608, needle position 3, idler jet 35, starter jet 90, idler air screw opened 1-2 turns
Gear ratios:	Crankshaft to countershaft 2,12, countershaft to mainshaft
	1st speed 3,62
	2nd speed 1,85
	3rd speed 1,24
	4th speed 0,86
	the same reverse

* CabrioLimousine

UNDERSTANDING AND TROUBLE-SHOOTING

ENGINE DOES NOT RUN

1. If your plug doesn't spark, but you get shocked by the plug wire, replace your bad spark plug.

2. Remove the "Black Box" cover. Check that voltage is present at the ignition coil terminal "15" using your test lamp. If it isn't, isolate the open circuit using your test lamp by first checking that voltage got from the battery to the ignition switch, then through the switch when it is turned on, and finally back through the wiring harness to the ignition coil.

3. Connect the test lamp between the engine ground and the coil's other terminal, "1". As the engine is turned over, the light should go on and off because the points, which are also tied to this terminal, are opening and closing intermittently shorting this point to ground. If this looks good and you have no spark, your ignition coil or spark plug wire is bad.

4. If the light doesn't go off and on, check your points under the front left engine cover (SA---). If they are not opening and closing, replace and - or adjust them. Connect your test lamp to the forward set of points on the right side of the cam, (or to the left set of points if you happen to be trouble-shooting the reverse ignition system now). If the points open and close but the test lamp never lights, voltage is not getting there from the ignition coil through the reverse relay.

5. Check the reverse relay. Voltage comes from ignition coil terminal No. 1 to terminal No. 1 on the reverse relay. Check this with your test lamp. The relay then connects this voltage to terminal "UV" for forward operation or to terminal "UR" for reverse operation. You can verify this with your test lamp while the ignition switch is turned on in either mode. Note how the reverse relay clicks loudly when reverse is selected. To correct Reverse Relay problems, see the REVERSE RELAY section below.

6. IGNITION TIMING. Timing is set by adjusting the points to open when the piston is a specified distance below the top of its travel. To measure this distance insert a depth gauge through the spark plug hole until it rests on top of the piston. Motorcycle shops sell these gauges, but you can improvise with a ruler or small diameter rod on which you can make accurate marks and measurements. Rotate the engine by turning the rear wheel while in second or third gear. Under the point cover, note that a black wire goes to the right, forward ignition points whereas a red wire goes to the left, reverse ignition points.

6a. Set the maximum contact gap at 0.4 millimeters (mm) to 0.5 mm (or approximately .016 inches to .020 in.) There is an eccentric and a set screw on each set of points.

6b. Set the timing by rotating the plate on which the points are mounted. This plate is secured by two set screws. Timing is set by rotating this plate so that the forward set of points opens (the test lamp goes on), when the piston is going up and is 4.5 mm to 5.5 mm (approximately .117 in. to .217 in.) before TDC. This happens to set the reverse set of points also. For reverse running, the points should open 3 mm to 4 mm (.118 in. to .157 in.) before TDC when the engine is rotated backwards. If you cannot adjust to these timing tolerances, your points or cam are worn too much and should be replaced.

7. Reconnect everything; the engine should now run.

ENGINE DOES NOT TURN OVER

(Try poking it in the ribs)

Your Messerschmitt runs well but the engine does not turn over with a good battery when the ignition switch is turned to the "start" position. If the engine turns over in either the forward or the reverse direction but not in both directions, the Dynastarter is good and the problem is in the Reverse Relay, so go to the REVERSE RELAY section.

1. Turn the ignition switch to the "start" position repeatedly, listening for the clicking of the starter relay in the Voltage Regulator. If you hear it, go on to step 4.

2. Check that the START signal is getting to the voltage regulator terminal "50" using the test lamp. If it doesn't, fix the wiring from terminal "50" on the ignition switch through the grey wire to terminal "50" on the Voltage Regulator.

3. Connect the test lamp to voltage regulator terminal "A". It should get power and the starter relay should click whenever power is applied to terminal "50". If not, either the contacts or the starter coil in the Voltage Regulator are bad. Disconnect the Voltage Regulator and remove its cover to determine the cause of the problem. The relay operation can be checked connecting the battery between the case and terminal "50". The contacts can be checked visually and with the battery and test lamp.

4. Connect the test lamp to the "Black Box" terminal block wires labled "A, 30h". It should light when the key is turned to "start". If not, check the wiring and return to step 3. If it lights, yet the engine doesn't turn over, check that there is a good ground connection between the battery, the engine, and the voltage regulator case. (Because it takes special tools and a lot of extra time to check the brushes and Dyna-starter, it is smart to take the next four steps to be certain the trouble is in the engine rather than in the "Black Box".)

5. Remove the engine harness from the "Black Box" lower terminals. All wire terminals are clearly stamped with the markings shown on the schematic. There are 5 large black wires, 2 small black wires, and a small red wire.

6. Connect the large black wires labled "B1" and "M" to the battery negative terminal (ground).

7. Connect terminal "B2" to "HE".

8. Momentarily connect terminal "A" to the 12 volt battery positive terminal. The engine should turn over. If it doesn't, the trouble is in the engine and the engine must be removed. Go to the DYNA-STARTER section. If the engine turned over, the trouble was in the "Black Box", so go to the next step.

9. Reconnect only terminal "A" from the motor to the "Black Box". Turn the key to the "start" position. If it turns over, it means the ignition switch, voltage regulator starter relay and motor brushes and starter are good!

10. Reconnect terminals "HE" and "B2" from the motor to the "Black Box". Turn the key to "start". If it doesn't turn over now, the Reverse Relay contacts are dirty or bad, so go to the REVERSE RELAY section. If it turns over, go to the next step.

11. Reconnect the motor ground "M" and terminal "B1" from the motor to the "Black Box". Turn the key to "start". If it doesn't turn over, the Reverse Relay contacts are dirty or burned open, so go to the REVERSE RELAY section. If it turns over, the engine will now start, so reconnect all the "Black Box" to engine wires.

THE MESSERSCHMITT ELECTRICAL SYSTEM

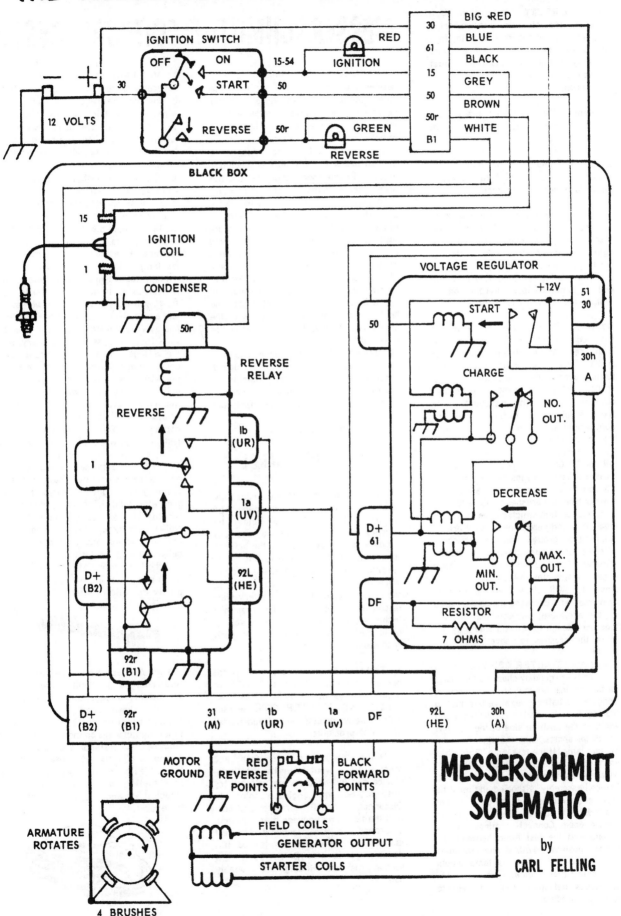

MESSERSCHMITT SCHEMATIC

by CARL FELLING

UNDERSTANDING AND TROUBLE-SHOOTING THE

MESSERSCHMITT ELECTRICAL SYSTEM

REVERSE RELAY If in any of the above tests the contacts were not connecting properly, remove the Reverse Relay from the Black Box noting the wire colors, markings and terminations. If you forget, the schematic is all you need to replace them properly.

1. Remove the cover. Note how you can push the central solenoid shaft to open and close all the contacts.

2. Examine and clean the contacts. The stationary arms can be bent slightly to allow good electrical contact.

3. Use a battery and two wires between terminal "50r" and the Reverse Relay case to operate the solenoid. Use your test light and battery across each contact to verify good electrical connections as the solenoid is being operated. Any problems will become obvious. Recondition or replace the relay if necessary.

DYNA-STARTER. The armature is mounted on the right side of the crankshaft. Most engines have armatures that spin inside the windings, but this one turns around the outside of the windings. So it also serves as a flywheel. A novel idea but with a maintenance disadvantage.--In order to gain access to the brushes, and field windings, it is necessary to remove the engine, the fan, and then extract the flywheel-armature with a special tool.

BRUSHES. Clean the area around the four carbon brushes and inside the armature. Oil from the crankshaft oil seal and carbon powder from the brushes combine to form a conductive film which causes shorts, so clean well and verify the brushes are insulated from ground. This can be checked with the test lamp and battery or an Ohmmeter. The brushes should not have excessive side play in their brass holders since this will allow them to turn in "reverse" and decrease the contact area on the armature. Squeeze their holders together enough to prevent side play yet allow the brushes to move in and out freely.

FIELD AND STARTER COILS Verify that these coils are good by checking for shorts to the engine ground or open circuits. Connect the battery negative terminal to the engine casing and one of the test lamp wires to the positive side of the battery. Check for shorts using the other wire on the test lamp. If the test lamp lights when you touch either "DF", "HE", or the "A" wires, it means there is a short to ground. The coil assembly is shorted and must be repaired or replaced. If the coils are not shorted, next check that they aren't burned open. Connect terminal "A" to ground and the test lamp between the battery positive terminal and terminals "DF", then "HE". The test lamp should light both times, indicating the coils are continuous, not open. Repair or replace open coil windings.

ARMATURE. Verify that the armature is neither shorted to ground nor open circuited. Reinstall the armature, and connect the test lamp between first wire "B1", then "B2" and the positive 12 volt terminal while the negative terminal is tied to engine ground. If the lamp lights, either the brushes or the armature is shorted to ground. Double-check that the brushes aren't shorted to their holders and ground, otherwise the armature is shorted to the crankshaft and must be repaired or replaced. Next check that the armature isn't open by connecting "B1" to ground and "B2" to one end of the test lamp while the other end is connected to plus 12 volts. The lamp should light indicating that current could flow through the brushes and the armature. If it did not light, the brushes were not seating properly or the armature is open and must be repaired or replaced.

SUMMARY. The Dyna-starter has now been checked out for shorts and opens. These indicate 99% of the problems, so assume it is now good. Reconnect everything; the engine should start now.

GENERATOR DOES NOT CHARGE

Your Messerschmitt now runs and electrically starts in both direction. This means that the Reverse Relay, the brushes, and the armature are good. But, the generator doesn't charge the battery: the ignition light stays lit when the engine is running; the headlights do not brighten when the engine RPM is increased; and the engine would die of the battery were disconnected. This means the problem is in the field winding not generating voltage or in the voltage regulator not properly regulating this voltage.

1. Test the field coils for shorts and opens between terminals "DF" and "HE" as described in the above DYNA-STARTER: FIELD AND STARTER COIL section. This usually detects any Dyna-starter generator problems.

2. If the coils test good for opens and short and you want to verify the Dyna-starter can indeed generate voltage, try this: a. Get the engine running. b. Disconnect the "HE" terminal and connect the test lamp between it and ground. c. Ground the "DF" terminal. The test lamp will light if the generator is good. If it doesn't go to the DYNA-STARTER section; the field windings are probably bad.

3. VOLTAGE REGULATOR If the generator is good, the charging problem is in the voltage regulator. Remove the wires, the regulator, and its cover.

3a. Manually operate the relays checking for smooth operation, and clean contacts to make good connections. Check that the resistor is good.

3b. Connect the test lamp between +12 volts on the battery and the "DF"-terminal. It should light brightly. When the over-voltage relay is manually depressed, the light should dim but remain lit. Next, connect the test lamp to the "D+, 61" terminal. It should light. Then connect the test lamp to the "51,30" terminal. It should not light, but when the undervoltage relay is depressed, the test lamp should light.

3c. If you have a variable power supply and a voltmeter, you can make the under and over-voltage adjustments of 12 and 15 volts (but this is seldom necessary, since the voltage regulator usually fails before going out of adjustment). If not, reconnect the wires but leave the the cover off the voltage regulator. (Warning: NEVER manually operate the voltage regulator relays when it is wired up! You could damage your Dyna-starter.) Connect the test lamp between ground and terminal "D+, 61". Start the engine and increase the RPM's. Watch the test lamp brighten and the relays operate. The fact that the lamp gets brighter verifys that the generator is good. The relays should try to regulate this voltage by limiting the field coil current through the "DF" connection and by opening and closing the circuit between the generator output on "D+,61" and the battery on "51,30". If you cannot clean and adjust the voltage regulator for proper operation, replace it.

Congratulations!

Your Messerschmitt now runs, starts, and charges in both directions! To thoroughly understand this electrical system, slowly read the theory and trouble-shooting sections again while following the wires on the schematic. With your new understanding of how to read wiring diagrams and use your test lamp, the lighting, or any other wiring for that matter, will be easy for you to troubleshoot and repair. Now, YOU'RE the expert.

The real thing: a dramatic photo of the original Messerschmitt KR 200 Super, complete with intrepid pilot on board

KRAFTWERKE

Here in the UK, a superb replica of the famous fifties Messerschmitt record breaker has recently come to light, as Peter Nunn relates

Early in the morning of August 29, 1955, a tiny, streamlined three-wheeled car was pushed out onto the tarmac of the Hockenheim race track in West Germany. Ahead lay a gruelling test of its capabilities – a 24 hour marathon involving six different drivers and the search for a new set of endurance records. Even at this inhospitable hour, a crowd had gathered to watch what was going on, the atmosphere being tense. The car, a special-bodied Messerschmitt, looked purposeful enough but how would it withstand the rigours of such an ordeal?

At the start, everything went to plan. Despite a heavy ground mist Bonsch, the first driver, set off punctually and was soon lapping with almost monotonous regularity. For hour after hour the Messerschmitt droned rhythmically round and round, the only trace of drama occurring when the windscreen became temporarily blocked with dead flies. Before the record attempt the authorities had

decreed that a 105kmh (65.2mph) average be maintained yet during its record-breaking run the Messerschmitt, although not quite equalling the 115kmh (71.5mph) set in practice, was able to average a comfortable 107kmph (66.5mph) with ease. Along the straights, moreover, it was pulling some 125kmh (77.7mph) while times taken at the end of one 30-lap spell showed a difference of only 1sec. The onset of foggy conditions in the last few hours of the run meant that the average fell to 95kmh (59mph) and that the positioning of blazing straw bales along the side of the track to mark the boundaries was necessary.

The team had certainly done well. Not only had 21 new endurance records been achieved but three existing targets set on the faster Monthléry track by a larger-engined car (350cc instead of the Messerschmitt's miniscule 191cc) had also been smashed. The record breaking run, which

Messerschmitt undertook to underline the sportiness, safety and robustness of the standard production car, had been a complete success. From a sales point of view, the fact that the record breaker was built with many standard parts, and had been driven with complete reliability for 24 hours was vitally important; they could, they said, have constructed a faster, more streamlined machine for the record attempt but there would have been no production parallel by adopting that approach. The record car, therefore, carried 60 kilos of ballast to simulate the weight of the base KR 200 model with two passengers on board, and was fitted with a production (albeit tuned) engine. The Dunlop tyres, drum brakes, headlamps and petrol were also of standard specification.

Epic record run

Joining drivers Bonsch, Strumm, Eisele, Rathjen and Dr Schwind in the epic record run was Fritz Fend, the indisputable father of the Messerschmitt cars. In the austere conditions of immediate post-war Germany, Fend began to experiment with simple pedal cars, tricycles and eventually ingenious conveyances for the disabled. The sheer lack of raw materials and natural resources, though, made his task incredibly difficult at times. By June 1948 the inventive Fend had constructed his first 'Flitzer' – a cigar-shaped contraption with an aircraft-type cabin, running on three wheels (two at the front, with a single driven one at the rear), and this in turn led to a succession of similar models and a meeting with Willy Messerschmitt in January 1952.

Between 1948 and 1952 Fend had improved the Flitzer quite considerably, to the point of making 10 cars per month in his own factory but difficulties arose when he tried to raise finance to expand. To cut a long story short (which is sad – Fend really deserves an article, or even a book, devoted purely to his fascinating exploits) Fend and Messerschmitt decided to merge their assets; Fend had plans for a new two-seater car and Messerschmitt was keen to branch out into other forms of mass-production as he was forbidden to build aircraft in his war-torn factories.

In the short term the old Flitzer remained in production but a new model code-named FK 150 – short for Fend-Kabinenroller 150 – powered by a Fichtel and Sachs 150cc engine, was announced officially in March 1953. Here was the first example of the familiar Messerschmitt outline, the new car having an opening plexi-glass hood, twin seats (one behind the other), three wheels and an air-cooled single-cylinder engine in the tail.

The KR 200, the basis of the Hockenheim record breaker, debuted in the Spring of 1955. In addition to the enlarged 191cc engine, the new car had hydraulic shock absorbers, a larger fuel tank, a wider front track, reverse gearing – in theory, it could go as fast backwards as forwards – and a proper accelerator. Top speed was quoted as 100kph (62mph) but this figure, we are told, was seldom

Early days: two Flitzer three-wheelers keep the streamlined record breaker company

'Gluckwunsch' or 'best wishes' for the victorious record crew and battered Super

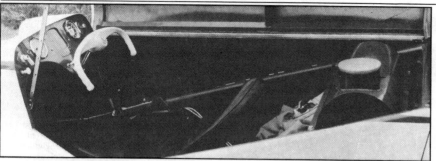

Top: a firewall will be added to the Super's spartan cockpit to cover the exposed petrol tank if Russell's proposed record attempt goes through. Right: Fichtel and Sachs 'single' plus lengthy exhaust tailpipe. Below: Russell Church and his remarkable creation. At the end of three years hard work, the car is a credit to his resourcefulness. Above: Fritz Fend seated in the emasculated Super original

reached. The final Messerschmitt variant, the fearsome four-wheeled Tiger, was introduced in September 1957, but in this highly-condensed pen portrait of the Messerschmitt company, there is only room to state that the Tiger, or Tg 500 as it should be called, was the peak of Fend's Kabinenroller theme. Even today a Tiger is considered fast for its size; in 1957 it was regarded as sensational.

After the record run, the car went on show in West Germany but ended up on display in the Deutsches Museum in Munich. During the 24 hour marathon, the bodywork had become battered in several places yet instead of restoring the car to its former glory, the Museum authorities inexplicably resprayed the paintwork, removed the nearside front wing, cut an 'access' hole in the engine cover and mounted the car vertically against a wall. Moreover the unique

'Super' logo on the side was left off, the engine fitted was incorrect for the age of the car (a post 1958 engine mysteriously appeared) and the KR suddenly acquired an extra gearchange mechanism when in its original guise there had only been one mounted on the left.

In the UK (and in Germany for that matter), one man knows probably more about this legendary car than any other, and that's Russell Church, a Messerschmitt enthusiast living in the Cotswolds. Not content with running a respectable KR 201 convertible or making spare parts for the Messerschmitt Owners Club, he's actually gone as far as to create an outstanding replica of the 200 Super. With the help of (principally) two friends, Russell took three years to complete this ambitious project with the aid of sketches, photographs and a

great deal of hard work. The monocoque which is of 1960 vintage, came from an associate in Bristol and, along with every other panel in the car's construction (there's no glass-fibre), had to be altered to the correct 1955 specification as original parts from around that time are simply not available. The engine, nose section, boot assembly, running gear and lifting section are all genuine Messerschmitt parts.

The attention to detail inherent in the Church Super is first rate, Russell going to inordinate lengths to ensure that the engine cover, for example, has eight slots instead of the customary seven and that the perspex windscreen (which was very difficult to make) matches the original perfectly. It should be stressed at this point that the entire project was a very home-made affair, there being no extensive professional aid given apart from that of panel beater Jim Keneston. Undoubtedly the hardest part of the project concerned the front wings, the first of which took at least 50 hours to construct. With the aid of cardboard patterns, a vast quantity of marine plywood and not a little skill, two wings were slowly and painstakingly constructed and then matched together on the car.

Mysterious box on bulkhead

Two anecdotes emphasise the difficulties Russell went through when building his Super replica. "I was looking at a photo of the engine bay of the original Super as it is now in the Munich museum" he relates, "when I noticed this small silver box mounted on the engine bulkhead. I'd never seen one like it before and no-one in the Messerschmitt Club knew what it did. That one really had me going until a friend who had been across to inspect the car at close quarters reported back that it was a transformer or something for the strip light which the museum were using to light up the engine bay! Little things like that were exasperating . . ." The other story concerns the cowled windscreen that Russell and helper Andy Carter produced for the Super replica. "We went to incredible lengths to get that right so imagine my despair when we finally had it fitted to the car; sitting in the seat, you just couldn't see through the thing clearly at all. I was reassured though when I heard that when sitting in the seat of the Munich Super you couldn't see anything through that either!"

"I viewed the whole idea as a challenge. It's not cost me a huge amount to build but without the assistance of my friends the car would definitely have been a non-starter. To build the car professionally would have cost a fortune."

Russell reckons his Super is good for 65mph flat out although with some subtle tuning it may just top 70mph. "It's pretty twitchy overall and not very practical to drive. If it rains, it just fills up with water!"

Why did he do it? Well, one reason is that he would like to have a crack at the records set by the original car 27 years ago. But do these still stand? Neither Russell nor anyone in Germany seem to know. As you read this, Russell is trying to locate a suitable venue for his own record-run (see separate news story on page 5).

Road impressions? Having spent a hair-raising afternoon sandwiched beneath its all-enveloping bodywork, I can confirm that the Super is terrific fun to drive! The steering is alarmingly direct and the forward vision akin to wearing swimming goggles – but the handling is respectable, the bike gearchange precise and the Super really moves. Or at least it feels as if it does. 50mph along the straight is fairly dramatic but 66mph averaged over 24 hours would be stirring stuff indeed.

In closing, it's slightly ironic that, from several aspects, the Russell Church Super is now 'more original than the original', to quote a popular phrase. Remember, too, that the car was constructed in a Gloucestershire loft without sight of the machine whose lines it reproduces so faithfully. Yes, that's right, Russell has never seen the KR 200 that inspired the idea in the first place. Nevertheless the accuracy of the finished result is very impressive. The question now is, will a Flitzer reconstruction happen? I think we should be told.

Achtung! Messerschmitt!

'When I pull up next to a Rolls-Royce or a Lamborghini I don't have to look at them; I know damn well they're looking at me.' So says the owner of one of about 40 Messerschmitts on Australian roads. And he's probably right . . .

WHAT DOES the name Messerschmitt mean to you? Baddies in Biggles books and Reach for the Sky, right?

"Blue Six to Blue Leader, bandits, bandits, six o'clock high."

"Blue Leader to Blue section, break to starboard, break to starboard."

"Roger, Blue Leader . . ."

You remember all that stuff. Well, that was a long time ago, and the Germans are our friends now. The wicked old Messerschmitt company has locked into the European aerospace industry on NATO's side, so its products aren't likely to be blackening the skies over London again in the foreseeable future. Just as well for all concerned.

What Messerschmitts do you remember? The Me 109 of course. The Me 110 twin-engined two-seater with the long glasshouse and twin rudders. The little Me 163 rocket plane and the Me 262 jet. What about the KR 200? You don't recall that one? Yes, of course it was a Messerschmitt. Come on, light transport type, two-seater, low altitude, unarmed, rear engine . . . ah, you've got it. You sneaked a look at the pictures.

It looks a little like the Me 163, with the rocket's little control surfaces replaced by a fixed undercarriage. Of course its performance isn't as impressive. Top speed is about 800 km/h slower and it has a very poor rate of climb. On the other hand it has better endurance and it's less likely to incinerate in a small fraction of a second if anything suddenly goes wrong.

You understand all this better when you learn that it really isn't a Messerschmitt design at all. Willy Messerschmitt's factory empire had expanded greatly in the late '30s, but most of it had been flattened by British and American bombers in the early '40s. After the war

the Germans weren't allowed to build aircraft, and the shrunken and struggling Messerschmitt company was reduced to making electric sewing machines.

In 1952 Fritz Fend approached the Messerschmitt management. He was an aeronautical engineer who'd served as a Luftwaffe technical officer. After the war he'd started his own small engineering shop, and designed and built some hand-operated three-wheeler invalid carriages. Then he'd done well with the powered version using a 38 cm³ motor-bicycle engine. He'd made a new motor carriage with a 98 cm³ engine in the body like a motor-cycle's sidecar, and this basic single-seater had been bought by fit people too, who were victims of the postwar car drought.

Some of these buyers had persuaded Fend to fit a scooter-type saddle and footrests for a passenger on the tail. Fend realised he might sell great numbers of a proper two-seater with a weatherproof top, but he didn't have the money or the plant space to try. He thought Messerschmitt might.

At Messerschmitt the management thought it wasn't a bad idea. They were planning to reopen their old works at Regensberg to build Italian Vespa motor scooters under licence, and they could fit a production line for Fend's new vehicle into the same program. They told Fend to get cracking on the prototype.

The prototype was basically the little Messerschmitt we remember. The chassis was steel tubing shaping a pressed steel tub, low in the front and high at the back, forming the floor and lower sides of the passenger space. The upper body was pressed sheet fore and aft, with a huge section amidships roofed with plexiglass and hinged on the right to open and let the driver and passenger in.

The front mudguards reminded you of spat fairings on aircraft with fixed undercarriage. The suspension consisted of a transverse lower arm acting on a three-element rubber spring component in torsion. The single tail wheel had another spring element connected to the aluminium alloy chain case which also served as a trailing swing arm.

The power unit was a two-stroke single-cylinder Fichtel and Sachs motor with a capacity of 173 cm³, giving just 6.7 kW at 5250 rpm and 12.7 Nm of torque at 4000. It ran on a 1:25 oil/petrol mixture. The gearbox in unit with the engine had four forward speeds, including an overdrive top, but no reverse. There was a motor-scooter pattern clutch, worked by a lever at the driver's right hand. There was also motor-scooter style steering, with very direct response to downcurved handlebars which carried the motorcycle-type throttle on the left hand grip.

The Germans make new words by sticking old ones together. The word they coined for the Messerschmitt three-wheeler was "Kabinroller", which hardly seems to need translation. The first model was designated the KR 175, from the motor's displacement.

It found plenty of buyers. It wasn't a terrific load carrier, it wasn't terribly fast, and the controls were tricky to learn. But it was cheaper and presumably more reliable than a used car, and it was more comfortable than a motorcycle. It had just enough acceleration to be a very practical commuter car. Many Germans who bought it grudgingly came to like it.

Fend introduced improvements. The KR 200 came with a 191 cm³ engine for a little more than 7.5 kW. A foot pedal replaced the awkward hand clutch. A reversing system was fitted, where the driver switched off the engine, then

moved a lever to make the two-stroke run backwards, then restarted, then engaged first gear as a reverse, and maybe then, if he was exceptionally brave and silly, all the other gears as high-speed reverses. There was a bigger fuel tank, a better back seat and a new bubble roof.

The KR 200 was another success. Nearly 12,000 were built in 1955, the first year of production, and it stayed on the market until 1962. A special-bodied and specially fitted version, driven by Fend and five others, ran nonstop for 24 hours on the Hockenheim racing circuit in August, 1955, to set 24 world records in its class and the class above.

It seems the first Messerschmitt three-wheeler to reach Australia was a KR 175 imported by a Sydney man as early as 1954. In 1958 an Australian Messerschmitt representative was appointed, and batches of KR 200s started coming out. It wasn't all that long since the war, so the Messerschmitt name was a dubious sales asset. But they were cheap, handy, allegedly capable of covering 100 miles to the gallon, and they had great novelty appeal. You saw a few of them around. You could hardly miss them.

They still have their fans. One fan is Eric Vargas of Sydney, who owns three examples. He is a leading figure in a new Messerschmitt club in New South Wales.

Five members brought their cars to town for the photographs shown here. "I also know of another five cars that are actually registered and on the road," Eric says. "I know of a few more people around who've got cars that are not registered. So altogether around Sydney I know of about 20 people who could be in the club. There are also a few in Western Australia. Originally 227 three-wheelers came to Australia, and of them I know of about 35 to 40 still left.

"I got my first one at a little garage in Wollongong. One day I was riding past on my bike, and this place was open, and I saw this little contraption, this little car, sitting just inside. So I went back and had a closer look. It was a bit of a mess, but the owner wanted $250 for it.

"In the beginning I thought that was very expensive, because when I sat in it the seat went right down to the floor. But I wanted the car, so in the end I paid him. Since then I've bought another one for $500, and that was in much better condition. The seats were original and in good condition, so I'll probably be able to get away without touching them. I was lucky to get it at that price, but it's still a better buy than the first one was. It's in a garage at Dural now, because I haven't got any space for it at Sydney. It needs some work on the engine, but apart from that there shouldn't be any problem with it. There's no rust anywhere."

Eric's third Messerschmitt is a rare and

attractive specimen. It's a Tg 500 "Tiger" four-wheeler, developed by Fritz Fend after the Hockenheim record success and put on sale in 1958. This model looked like a standard Kabinroller with an extra back wheel, but there were a lot of changes under the new tail.

The four-wheeler's engine was a specially designed aircooled, twin-cylinder Fichtel and Sachs two-stroke with a displacement of 493 cm^3, running on a 1:40 oil and petrol mixture to give 14.5 kW at 5000 rpm. Third gear and top in the unsynchronised four-speed transmission were both overdrives. There was a normal reverse gear. The engine and transmission sat on a steel tube subframe attached to the Kabinroller-type monocoque. There were useful changes to the suspension. It was 15 cm longer than the three-wheeler, five cm wider, 135 kg heavier, and capable of 124 km/h.

The four-wheeler was offered with the three-wheeler's range of cockpit styles, bubble canopy, canvas "roadster" convertible top and short "sport" streamlined racing-type windshield. But in spite of fuel consumption figures of about 16.0 km/l (45 mpg), the "Tiger" sold to enthusiasts, not miserly commuters, and the open-cockpit "Sport" version became the definitive type. A 1959 example with a new streamlined nose was timed at 138 km/h on a series of two-way runs.

Eric says only about 320 four-wheeler

Messerschmitts were ever built. A leading German Messerschmitt enthusiast, Heiko Zimmerman, says 150 are left.

"Only five four-wheelers came to Australia," Eric says. "We know of four, and one of them is still missing. Some people modified their three-wheelers to make them into special four-wheelers. There's been quite a lot of modification of Messerschmitts over the years. One guy, he had a three-wheeler and he put a Triumph 1000 Tiger motor in the back of it. He reckoned he got it up to 160 km/h. One bloke in Britain I know of lifted the whole front end of a Mini into the back of his, and he hillclimbed it.

"I bought my Tiger off Green's Motorcade Museum and I paid $2700 for it. It goes like a little beauty but there are a few little problems with it. The brake linings and shock absorbers had to be replaced. Originally it had a vacuum-operated fuel pump on it and I'm not quite sure of the reliability of those, because every one I've seen so far has had that changed to an electric type. Mine's got an electric pump, but the trouble is there are two hoses coming from the fuel tanks, one from the main tank and one from the reserve tank, and you need a little tap to bring those together.

"There are a few little things like that, but otherwise it's in quite good nick. Once I've got those things done I'll take it down

This swarm (left, top) is 12.5 per cent of the Australian total. With canopy and engine covers open the KR200's unusual aspects are obvious. Single-cylinder 191cm³ two-stroke (left) is very simple. Twin-cylinder 500cm³ Tiger (above) isn't exactly high-tech either. Cockpit (top) is also basic

to the body workshop and get it sprayed, and she'll be right."

Fred Diwell of Kirrawee, a club member, has another four-wheeler that's roadworthy but not yet registered. When the first three-wheeler Kabinroller got to Sydney, the authorities scratched their heads, peeped at the engine, counted the wheels and eventually decided to tax it as a motor-scooter. But later on they had second thoughts about letting a helpless member of the public get off so lightly, and since then they have been taxed as cars.

Messerschmitt drivers can put up with that and set it off against easy parking and the fuel they save. "I do about 60 to 65 miles per gallon (21.3-23 km/l) in mine," Pat Marshall-Cormack says. "I can get 90 (32 km/l) if I try hard. I've done about 29,000 km in it in two years, which for a car with 200 cm³ motor is fair going. You get about 14,500 km out of a motor before something major goes wrong and you've got to pull it apart. I'm on my third now.

"There are a few problems like that. You used to have tyres that size made here and they were quite good, but now you have to use Korean-made tyres and they're not much good. They don't have a

good tread and they wear out too fast.

"It's the sort of car you've got to be dedicated to drive. Actually I find it's fun to drive, I think it suits my ego. When I pull up at traffic lights next to a Rolls-Royce or a Bentley or a Lamborghini, I don't have to look at them; I know damn well they're looking at me."

What are Messerschmitts like to drive? Well, they're interesting. Their performance varies a great deal from use to use, depending on how many hills get in the way. On a good level stretch they're not too bad. The busy sound of the engine and the sight of the ground zipping past just out of arm's reach might well convince you that you're fairly pelting along, until something much bigger suddenly fills your mirror and rocks you with its slipstream and shoots away ahead.

When a hill gets in the way, even this modest level of performance is seriously impaired. Then it's a case of plugging on up in low gear, and being patient until you're over the top. The engine's note is a cheerful ripping snarl on the level, but in a climb it falls off to a choking pop-pop. It's working awfully hard and you really feel for it. As a commuter car in the city, of course, these good and bad points wouldn't count nearly so much. You'd spend so much time standing still with everybody else at the lights, it would hardly matter what you were doing in between times.

The ride is something you'd have to get used to. The suspension is very stiff, good enough for European roads but not so good for the decayed surfaces around Australian suburbs. On a good surface it can whip round corners almost without needing to slow, but on uneven going it has to be handled very gently. Again, the very direct steering by handlebars and the wide turning circle are details you'd have to learn to live with. Not that the turning circle is bad by full-sized car standards, but you'd think they could have made it much smaller. Perhaps they didn't want to let drivers put its stability at risk.

The driver's view is excellent, although it might also be a bit unsettling. If things got out of hand, there wouldn't be much car between him and whatever he was going to hit. A Messerschmitt is outbulked by, say, a Honda Civic in the way that the Civic is outbulked by a double-decker bus. The bubble roof is pretty claustrophobic. This doesn't matter so much to the driver, who has other things on his mind, but it gets to the passenger and the little sliding windows are very little help.

Any motor vehicle is basically just a fast, handy and comfortable alternative to walking. In the case of the Messerschmitt, this is particularly obvious. As Germany's postwar economic recovery got under way, Volkswagens, BMWs and Mercedes came to look like even better alternatives

to walking, and Messerschmitt owners began to appear miserly, deprived and eccentric. Under these economic and social pressures the market for rock-bottom personal transport was shrinking.

In 1956 the Allied restrictions on the German aircraft industry were lifted, and the Messerschmitt company suddenly found a range of much more exciting projects open to them. Fritz Fend's enthusiasm helped keep the little Kabinroller line going a few more years, but the last "Tiger" came out in 1961 and the last three-wheeler was sold in 1962. In nine years about 50,000 had been made, not a bad record for an eccentric bargain-basement stopgap.

Messerschmitt sales always depended largely on the home market, and this was the most prestige-obsessed and power-minded in Europe. But in Germany and elsewhere, there were always a few people who were sensible enough to settle for three-wheelers to meet their wants, or eccentric enough to actually admire them. Despite their fragility and small size, a good few have survived. Now, they're becoming collector's items.

"In Australia we really haven't got much to judge by," Eric Vargas says. "There's not really a market, so you can only put a price on one and see if it sells or not. I know of one in Adelaide and one in Queensland that were sold a few months ago. I didn't see the cars but apparently they were well restored, and they asked four and a half thousand for one. I saw one at Green's Motorcade Museum, it was an orange three-wheeler and it was very tatty, it was very rough generally, and at the auction that one went for three and a half thousand.

"There was one interesting sale about a year and a half ago. Pat Marshall-Cormack had a white Messerschmitt that was good mechanically, but with the body and seats tatty and a bit run down, needing general restoration again. He put it in the paper one Saturday before he went overseas, and for an asking price he put two thousand on it. And by nine o'clock in the morning he'd had 37 phone calls, and he'd actually sold it."

So there you are. It looks as if the Messerschmitt will be a restorers' cult car, after the Morris Minor and the Bugeye Sprite and the FJ Holden, probably just ahead of the VW Beetle. More power to them. Seen, once again, strictly as an alternative to shoe leather, it's a bloody clever little machine. It's probably the sort of car we ought to have more of right now. Air-conditioned stereophonic living-rooms on wheels are a waste of time, money, petrol, rubber, plastic and anything else you can think of. As the oil burns away, the price goes up and the reserves go down, these will come into their own. Wait for it. □

MESSERSCHMITT
KR-200

by
J.F. Blondelet
Translated by **David Cullen**,
with special thanks to
Bruno Garcia of Nommay, France,
for his proofreading of the text. The
original French text article appeared
in **Auto-Rétro,** April, 1982.

Certain legends have a hard life. . . At the end of the 50's it wasn't rare to see Messerschmitts on the streets of Paris, and, like all my school buddies, I was fascinated by this odd little airplane that had lost its wings. The war wasn't very far behind us and many of us knew that the Luftwaffe had used Messerschmitts, the most well-known being the BF109. When we looked at a Messerschmitt car, it was like looking at an airplane cockpit furnished with wheels. It was only much later that I learned that the Messerschmitt Kabinenroller, although it looked like an airplane fuselage, had nothing to do with the legendary BF109, except that it carried the name of the great aeronautical engineer, Willy Messerschmitt.

The Birth of the Karo. . .

At the end of the Second World War, the Messserschmitt company had 45,000 engineers and qualified workers spread out in several factories that had been mostly destroyed by incessant bombing; knowing that Germany was no longer allowed to manufacture aircraft for an indefinite period, the tendency was toward firing rather than hiring, employees. In 1952, right in the middle of a restructuring crisis for Messerschmitt, Fritz Fend came on the scene. Fritz Fend was an engineer by profession, but an inventor by genius; he had created and produced a car for invalids, a sort of tricycle with two rear wheels and one front steerable wheel; it was powered by a moveable handlebar that operated forward-and-back. He improved the design, switching the wheel arrangement to one rear wheel and two front wheels for steering, put in a rear-mounted 38cc, two-stroke motor, and covered the whole thing with an aluminum cabin that resembled a side car. Although originally conceived as a car for invalids, this little vehicle caught the eye of a lot of people who were looking for cheap transportation in a Germany that was recovering from several years of war.

This little car, baptized Filtzer, was a one-seater; it sold for 1500 to 2000 marks and about 250 were built.

About that time (toward the end of 1951) Fend met Messerschmitt. Messerschmitt saw excellent possibilities for building the Flitzer in his factory at Regensburg; he asked Fend to rework the design of the Flitzer; he lengthened it so he could add a second seat for a passenger, equipped it with a bigger motor, and gave it a look just like a cockpit on wheels.

The contraption thus created was unique in its conception and (one might say) plainly inspired. The Kabinenroller came along at the right moment at a time when such a vehicle was desperately needed.

The first models were called **KR** (for **K**abine**n**roller) 175's—the motor was a 175cc Fichtel & Sachs. Well-equipped (speedometer, turn signals, outside mirror, good tool kit, windshield wipers, etc.) the Karo (as it was nicknamed) let its owner move from one place to another without getting dirty and with a certain amount of comfort; a good number of German industrialists made it to their first business meetings in Karos! In 1954, the motor was increased to 200cc and equipped with an electric starter: the new model was called the KR 200. This version, the most well-known, was built up until 1961, the date at which production was finally halted. [BN EDITOR NOTE: Production of all Messerschmitt models ceased in January of 1964.] Several models were available: the standard with bubble top, the Kabrio with side windows, the roadster with only a very small front windscreen; there was also a luxurious export version with a two-tone paint job, a nicer interior package (heater, clock, and a protective cover for the top in summer time) and a bit of extra chrome.

A competition model of the KR 200 even raced at Hockenheim for publicity purposes.

In 1956, when the production of aircraft in Germany was reinstituted, Messerschmitt and Fend went their own ways. Fend bought the factory at Regensburg and started a new company, FMR (**F**ahrzeug and **M**aschinenbau GMbh **R**egensburg), at the same time keeping the right to use the name Messerschmitt on his vehicles.

In his perpetual search for perfection, and following the success of the competition model of the KR, Fritz Fend crated an incredible machine in 1958: the FMR TG 500 (the 500 represents the displacement and the TG stands for Tiger). On the outside, it looked like a KR 200 with four wheels; it was in the mechanical end that the differences were important. The Fichtel & Sachs motor was two-cylinder, two-stroke, air-cooled, transversely mounted between the two back wheels. The motor was tied to a four-speed gear box with a genuine reverse (you will see in a bit why I say a "genuine" reverse!).

This four-wheeler had all the qualities of the three-wheeler, with the benefit of added power exceptional for a vehicle of this type (more than 75 mph)! Again, several versions were manufactured: convertible or with bubble, for a price around 3500 marks.

But, it was the end of the 1950's, and the German auto industry was back on track; Volkswagen was putting out beetles by the millions. There was no longer a place for the little cars, and the Kabinenroller and the FMR TG 500, although 500 units were built, were never fully exploited. Born out of the wedding of a genius inventor and an aeronautical giant, the KR 175 and 200 ran through a production of about 50000 units (all models and options considered); a great success for a product conceived and built at the exact moment of its need. . .a perfect answer for a question never asked.

The Engineering of the KR 200

Actually, the principal frame member of the Messerschmitt looks a lot like the central part of a modern Formula One, with a tubular brace supporting the sides and a stamped sheet metal deck to increase the front drive, a transversal arm supporting the wheels and the steering mechanism; on the rear is the drive mechanism, looking a lot like that of a scooter with the wheel mounted on the oscillating arm, and the transmission running off of a covered chain drive.

"KABINENROLLER"

The Messerschmitt KR 200 in 1982

The somewhat bulbous body is riveted on the frame box with two little fenders covering the front wheels, a hinged hood forming the whole rear section, and of course, the famous and characteristic bubble top, also mounted on hinges and able to be swung aside completely to permit passengers to enter.

The driver's seat is mounted in such a way that it moves up and back so that the driver can get in; a small bench seat in the rear allows an adult and a child to seat themselves with a bit of comfort. The "cockpit" has a handlebar for steering, a speedometer, a very small shift lever, and a standard pedal arrangement; there are also a hand brake and the switches for the headlights, the horn, windshield wipers, and turn signals.

How It Handles

Within a few feet you get used to the very direct steering; the least movement to the left or right on the handlebar, and the Messerschmitt literally leaps in the desired direction after zigzagging for several hundred feet; you get used to it! It shifts like a motorcycle, 1st at the top, 2nd, 3rd, and 4th by pushing the shifter toward the bottom; as for reverse. . . you must stop the motor completely, switch the motor into the opposite direction—which is possible on a two-stroke—and put it into 1st again. If you're brave or if you're a kamikaze, you can go through all four gears in reverse! The brakes are activated by cable and don't pose any particular problems. With a bit of experience, and thanks to the narrowness of the body, it's possible to thread through city traffic with lots of agility and dash; the motor has a lot of flexibility, but lacks pick-up just a bit. On the other hand, the steering, even if it's alright at road speed, has some problems at low speed on account of its too large turning circle. Road handling is fabulous and it is almost possible to turn at right angles with just a light touch on the handlebar in the direction desired. Top speed is about 50 mph—not bad, given the displacement. Visibility is excellent, of course, and comfort is very acceptable. Maintenance is as easy as with a motorcycle.

We asked several questions of Fabian Sabatès, who was kind enough to lend us his Karo for this article.

Fabian, what's it like driving a Messerschmitt in Paris in 1982?

You get the impression from the driving public that you're an idiot; they really hold you in contempt, as do the police. Even the oddball at the wheel of his "metallic violet Capri with fuzzy steering wheel cover and baby shoes hanging from the mirror" hates you. People neglect you, they don't see you, and they refuse to yield the right-of-way. Sometimes you make them laugh because they're asking themselves what that refugee from a merry-go-around is doing on the road! You get their pity because they think it's a cart for invalids; sometimes you get their sympathy. Kids really like it, because it's on their level and built on their scale.

When you're under the bubble, what's it like?

First of all, the Messerschmitt does nothing for your love life; you can get a stiff neck trying anything in a Messerschmitt (ed.: along these lines, nothing can beat a 1949 Nash Ambassador back seat for intimacy!). It's the only car in which you are unable to read a map because of the space problem. It is, however, well-designed inside, and there is no sensation of claustrophobia; on the other hand, in the summer, the heat is awful—that's why a cover is so useful.

What can you tell those who want to buy and maintain a Messerschmitt?

It's a rare vehicle. Many have been scrapped, but the principal enemy is rust and that is the first thing you have to get under control when you find one. Finding parts isn't too difficult, but it's best to get one in complete condition. Germany and England have clubs that remanufacturing parts, like the bubbles for instance, but the wait can be long because of the minimum number of orders necessary to do a manufacturing run.

Why did you buy a Messerschmitt?

To make a dream come true; probably the same dream a lot of kids have: to finally own one of these odd little cars with a frog's face, like the ones that I liked so much as a boy. . . **Santa Claus must have heard your request!**

Specifications

Motor: Sachs, type 200 LDR, one-cylinder, two-stroke, Schnürle brushes. Bore 65mm, stroke 58mm (2.56 x 2.28 in.); a motor that is well-balanced, with a stroke/bore ratio of .89. Displacement is 191cc, compression ratio is 6.3-to-1.9.7 hp at 5000 rpm. Air-cooled.

Drive: Motor and four speed box. Primary drive is 2.12-to-1. **Internal ratios:** .86 (100%), 1.24 (69.3%), 1.85 (46.5%), and 3.62 (23.8%)-to-1. **Secondary chain drive:** 13/30 teeth. Reduction ratio of secondary is 2.3-to-1. **Overall ratios:** 4.16-, 6.04-, 9.01, 17.63-to-1.

Ignition: 12 volt generator, 90-135 W furnished with two sets of points, each on a separate circuit, which lets you reverse the motor to put the car in reverse by means of a two-position switch. **Angle of advance:** 4.5-5.5 mm before top dead center in forward gears; 3.4 mm before TDC in reverse.

Plugs: 225; 240 for sport use.

Carburetor: Bing choke-type 1/24/87 or 1/24/88. Main jet 120; needle jet 1608, needle at the 3rd notch. Idle jet 35, choke jet 90. Idle jet screw backed out about 1-½ turns.

Frame: Triangular tube-type and body, stamped steel frame, ribbed for rigidity.
Front suspension: independent wheels, rubber torsion springs, double-action hydraulic shocks.
Rear suspension: oscillating also furnished with rubber torsion springs and a hydraulic shock.
Wheels: 4.40 x 8, drum brakes on all three wheels run by cables. Gas tank holds 14 liters of gas-oil, of which 2.5 are the reserve.

Dimensions: length-2820 mm; width-1220 mm; height 1200 mm; wheelbase-2030 mm; front track 1080 mm; clearance-160 mm. Weight: 506 pounds.

the airplane of the road

Red Baron, Your Car's Ready

by Tim Howley
photos by the author

M ENTION the name "Messerschmitt" in collector car circles, and the conversation immediately turns to, "Oh, yes, those skittish little cars designed and built by that German aircraft company." You'll hear stories that Messerschmitts are tricky to drive, can be driven as fast in reverse as in forward, and get 100 mpg. Since the vehicle is just beginning to receive collector attention, it is understandable why the misconceptions still exist. Many of the highly colored stories were actually founded on misleading factory and distributor advertising and on less than completely informed reports in the motoring press of the time.

The truth is that Messerschmitt did little more than lend its name and production facilities to somebody else's idea. A Messerschmitt can be as safe to drive as any contemporary sub-compact, providing you fully understand it. Anybody who ever tried driving one at highway speeds in reverse is probably no longer here to tell about it. To say it gets 100 mpg is like trying to compare your everyday driving habits to the Mobilgas economy runs. To call it a car, or anything else, is to completely misunderstand it. The Messerschmitt combined features of a small car, motor scooter and airplane in a very special way. You might say the Messerschmitt was the most unusual vehicle to come down the nifty roads of the fifties, and that would probably be the most accurate of all descriptions.

Its namesake, Messerschmitt, began in Bamberg, Germany, in 1923, at first constructing sailplanes and sports aircraft. Then came the M-18 in 1926. The famous Bf 109 was developed in 1934-35. Later came the Me 262 Sturmvogel and the Me 163 Komet, among the first operational jet and rocket aircraft of the war.

When the war was over, Messerschmitt's main factory at Augsburg was 75 percent destroyed by bombing. Moreover, Germany was prohibited from building aircraft. Thousands of Messerschmitt workers were unemployed, in-

cluding engineers and skilled craftsmen. There was little that founder Willy Messerschmitt could do about it.

In 1952 Fritz Fend arrived at Messerschmitt with a plan to get the Regensburg works back into production. Fend was both an idealist and an aeronautical engineer. He had been a valuable technical officer in the Luftwaffe, but had little business sense.

In the late forties, Fend produced a few three-wheeled scooters for invalids. The first ones were hand-operated, but later versions had small motorbike engines. A third version, called the "Fend Flitzer" became quite popular. Though still intended for invalids, many Germans bought it in lieu of new automobiles. Remember, this was a period when the German economy was devastated. Few Germans could afford new cars, and the prices of prewar cars were unreal.

The Fend Flitzer answered a lot of needs. It had a two-stroke, single-cylinder engine, unitized tubular steel frame and aluminum sheeting. It sold for approximately $300. By 1951 both an open and closed version were offered.

You mounted either vehicle by tilting the entire cockpit forward. Fend also produced a three-wheeled cargo carrier. Unfortunately, Fend had neither the business acumen or financing to realize the full potential of his vehicles. But he did have excellent connections in the German aircraft industry, and approached Willy Messerschmitt. What he proposed was a stretched Flitzer with a more powerful engine. Messerschmitt was already planning to produce motor scooters under license from Vespa, had the financing to back a new vehicle and was eager to get its Regensburg plant back into production.

It is still wrongly believed that Willy Messerschmitt designed and built the namesake car, and that it was purposely made to look like the Pf 109. Rather, it was Fend who developed the Messerschmitt in 1952 and 1953, inspired by his own Fend Flitzer. Willy Messerschmitt had very little to do with the project.

Granted the Messerschmitt looked like a World War II fighter plane. Its three-wheeled design, two wheels in the front and one in the rear, inspired

drive report

1955 MESSERSCHMITT KR-200

MESSERSCHMITT
continued

tandem seating in true small-aircraft tradition. The handle bar, if turned upside down, could have been placed in an aircraft cockpit. The bubble dome had a decided aircraft appearance. But all of these features and more were originally designed for function and to keep the price down rather than to make the vehicle look like something it wasn't.

The "Kabinenroller," as it came to be called (meaning cabin scooter), was a happy compromise between a motor scooter and a light automobile.

The KR-175 was introduced in 1953. The initials "KR" stood for Kabinenroller and "175" for the displacement, which was actually 173 cc. Wheelbase was 79.9 inches and overall length was 111 inches. Front track was 36.25 inches. The vehicle was 48 inches wide and 47.2 inches high. Its basic uniqueness stemmed from the frame structure. The duplex tubular frame of welded-up steel construction carries the front suspension, then goes diagonally upward to hold the struts that mount the rear suspension. The entire body is slung within this framework. At the rear is attached a separate forked sub-frame which carries the engine. The sheet-metal body panels are riveted to the frame. There is even a sheet-metal belly pan under the frame. Since there is no chassis under the vehicle, hundreds of pounds of weight are saved, enabling the maker to build a complete vehicle weighing only 480 pounds. Rubber torsion suspension is independent at all three wheels, but lacks hydraulic shock absorbers in the 175 models. Steering is the motorcycle type, with chromed handlebars linked up directly with the front wheels. Steering is unbelievably quick, and takes some getting used to. To further complicate the operator's control, a twist-grip throttle is placed on the left handle bar. Mechanical braking is accomplished through a conventional foot pedal plus a parking brake lever on the inner right side. Front wheels are housed in aircraft type fenders without wheel cutouts.

Power is supplied by a blower-fan cooled, two-stroke, single-cylinder Fichtel and Sachs engine of 173 cc which develops nine horsepower at 5250 rpm. The final drive to the rear wheel is a motorcycle chain. This kind of a setup eliminates the extra expense and weight of the usual driveshaft, universal joints, differential assembly and rear axles. But it results in a transmission with some real peculiarities, as we shall soon see.

The four-speed unsynchronized transmission is operated through a gear lever to the right of the handlebar.

Above: Messerschmitts have been described as airplane cockpits on three wheels, and the comparison is apt. Right: Fichtel and Sachs two-stroke engine can power the little car to speeds just a tad over 60 mph.

A trigger type "neutral select lever" is mounted in the gearshift to provide a neutral position between all four gears, meaning that in heavy traffic you have the convenience of being able to take off in any gear without going all the way back to first. No floor clutch is provided on the KR-175. Actual shifting is done by a special cable that shifts gears by a push-pull action. In practice, you pull back into the lower gears and push forward into the higher gears. The real problem with the 175 transmission is that it has no reverse, primarily because the manufacturer wanted to hold down production costs. If you want to reverse the car you get out and push.

The top section is a rather unattractive squared-off bubble dome of plexiglass. There are sliding plexiglass windows in aluminum frames on both sides and there is a flat glass windshield in front. The windshield wiper is hand operated. To enter the car, you lift up the entire top section. The body is cut down in front so access is quite easy. You enter from the left side. The top section is hinged on the right side and a

strap keeps it from accidentally blowing all the way open and cracking the bubble top on the ground. The tail section, which houses the engine, is hinged from the front, allowing complete engine access. The vehicle looks quite bizarre with both the top and engine compartment open.

For 1955, the KR-175 was followed by the KR-200, both because of the former's sales success and in answer to the many complaints about the vehicle's peculiar characteristics. The number "200" stood for the 200 L Fichtel and Sachs engine of 191 cc, giving it 10.2 horsepower at 5250 rpm. A complete set of foot pedals was now offered for clutch, brake and throttle, but the peculiar ratchet type gearshift lever with its neutral trigger remained.

At last there was a reverse, and what a crazy way to go in reverse. You shut off the engine, push the ignition switch in and turn the key on again. A green light below the red generator light now lets you know that all gears are in reverse. The reverse sequence is accomplished through an electrical system that is as

unique as every other Messerschmitt feature. When you depress the key you cause the relays at the back of the car to reverse the windings going to the starter. At the same time another set of points is switched, allowing the ignition to fire just ahead of top dead center. When this happens to a two-stroke engine it actually runs backwards, yet charges the system normally.

Rumors quickly spread that you could drive the new Messerschmitt KR-200 faster in reverse than in forward. Not true. Servicemen in Germany actually set out to prove it by driving the little vehicles on air strips. But a steering system that was merely "touchy" going ahead became deadly in reverse, and at any speed the vehicle could easily overturn with no more protection for the driver than the plastic bubble top.

Other major improvements in the KR-200 included raising the fourth and final drive ratios, reworking the rubber torsion suspension somewhat and adding hydraulic shocks on all three wheels. For improved stability, the frontal track area was increased from 36.25 inches to 42.5 inches. The 175's awkward looking plexiglass top was replaced with a more rounded bubble top which had a curved glass windshield and an electric windshield wiper. One big improvement was

front wheel cutouts which were the result of owner complaints of having their fenders dinged in parking lots. Some of the minor improvements included a plastic steering handle, two-piece rear seat, larger and more accessible gas tank, parking brake lever on the left side instead of the right, small eyelids adorning the twin headlamps and side blinkers moved from the front to the rear. All in all, the KR-200 was a much more stylish and manageable vehicle.

Management must have made a lot of right decisions, because in the 1955 introductory year 11,909 units were

Tomorrow's Trikes

Small three-wheelers are by no means relics of the past. There is now the H-M (High Mileage Vehicle) made in Minnesota, the Columbia and the BD-200 slated to be produced in Wisconsin. H-M has two front wheels and one rear wheel, seats one, and is supposed to get around 80 mpg.

The "Cub" is produced in Taiwan by Convenient Machines of New York. The Cub has two wheels in the rear and one in the front. It weighs only 600 pounds and features aerodynamic styling with a fiberglass body strengthened with a welded steel frame and built-in roll bar. It has a four-cycle 300 cc Honda-built engine, automatic transmission, electronic ignition and rack and pinion steering. It has full opening side doors and a lockable hatchback.

The most talked about one of all is Ford's Ghia Cockpit (said to be a KR-200 look-alike) introduced at the Geneva Auto Show in 1981. It has been traveling the international show circuit ever since. It has two front wheels, one rear wheel and tandem seating. Designed pretty much along the same lines as the Messerschmitt, it was developed by Ford's Ghia Operations of Turin, Italy. It is an extremely aerodynamic design with tubular chassis, 12 horsepower, 200 cc air-cooled engine and electric reverse gear. The clear plastic canopy is hinged from the front rather than from the side. There is a steering wheel instead of a handle bar, and instrumentation and controls are far more conventional than with the Messerschmitt.

Ford has no immediate plans to produce the vehicle but is currently testing public reaction to it. Ford says the Ghia Cockpit combines tremendous fuel economy (in the 95 mpg range) with more than adequate performance and maximum speed for today's driving conditions. However, in its present form the Cockpit does not meet all aspects of current safety legislation in this country, and for that reason alone may not see production in the near future.

MESSERSCHMITT

continued

sold, establishing a one-year sales record. The new Kabinenrollers were promoted heavily with a record run in the fall of 1955 on the Hockenheim circuit. Messerschmitt set 22 international records on a 24-hour run, convincing Fend more than ever that he had produced the ultimate vehicle for the postwar world, and all of Europe would be flocking to his door. But, as we said earlier, Fend was far more of an idealist than an industrialist. First, there was now German competition from BMW Isetta (see *SIA #70*) and Heinkel. Second, in 1956 the ban on German aircraft construction was lifted and Messerschmitt now had bigger fish to fry.

Messerschmitt sold the Regensburg works, and Fend organized a new company called Fahrzeug-und Maschinenbau GmbH, Regensburg. The marque officially became the FMR from the firm's initials. The production of both the Messerschmitt 200 under the new name and the three-wheeled cargo carriers continued.

By 1957 four different versions were being offered. In addition to the Cabin Scooter there was the Cabriolet or 200 Kabrio with framed side windows and a folding fabric top. Then there was the KR-201 roadster with a fixed safety glass windshield, side curtains and a folding cloth top. Finally there was the Sport, a stripped version with a low plexiglass windshield and no top, just a removable cover extending over both seats.

The final and finest Messerschmitt came in 1958. This was the FMR Tg-500, originally called the Tiger. At first glance it looked like a KR-200 with two rear wheels. In actuality only the nose section, front suspension and body-frame were taken from the earlier vehicle. These were some of the improvements: A more refined swing-axle type rear suspension was now employed. A transverse engine/transmission was now added. The two-cylinder engine produced 19.5 horsepower at somewhat of a sacrifice in fuel efficiency. Finally the shift lever was given a standard four-speed gate plus a reverse. Clearly, the once totally unique Messerschmitt was moving closer and closer to becoming a conventional automobile, if not in appearance, certainly in engineering.

The FMR Tg-500 was offered in a coupe and a roadster and later in a sport version. In 1959 a convertible top was made available to fit over the side frames as on the KR-200 Kabrio.

Only a few hundred Tg-500 models were built before production ceased in 1961. A year later production of the KR-200 (now called the 201) was

Above: Entry and exit is gained by lifting entire roof structure and stepping over left side of the little fellow. Right: Final drive from four speed transmission is by chain running in an oil bath. Below: Original radio mounts in center of dash.

stopped. Approximately 50,000 Kabinenrollers were built during the nine-year production run, of which only about 10,000 were 175s. The 200 was the most predominant model. Some 300 Kabinenrollers of all types are known to survive today in this country. Any Messerschmitt you find today is quite rare and probably worthy of restoration. Compared to automobiles they are relatively inexpensive to restore and are quite valuable when completed.

Driving Impressions

Our driveReport car is an all original 1955 Messerschmitt KR-200 owned by Carl and Marilyn Felling of Topanga, California. Marilyn Felling is co-founder and president of the growing Heinkel-Messerschmitt-Isetta Club and is editor of its enthusiastic publication, *Bubble Notes*. Carl and Marilyn own at least two dozen mini cars of the mid fifties.

They found this Messerschmitt about seven years ago in Los Angeles. Back in 1960, the original owner had given up on the car because his local dealer had gone out of business. Rather than sell it at a depressed price, he decided to wrap it up in a plastic bag and hang it by the rear wheel from the side of his garage. There it sat until the Fellings heard about it some 15 years later. It is very

specifications

Illustrations by Russell von Sauers, The Graphic Automobile Studio

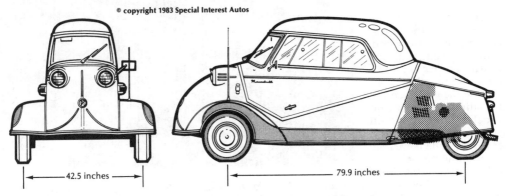

42.5 inches

79.9 inches

1955 Messerschmitt KR-200

Base price 2535 Deutschmarks

ENGINE
Type	Fichtel & Sachs single cylinder, two-stroke, air cooled, ported.
Bore and stroke	2.56 x 2.28 inches
Displacement	191 cc (11.7 c.i.d.)
Max bhp @ rpm	10.2 @ 5250
Max torque @ rpm	11 lb./ft. @ 3800
Compression ratio	6.6:1
Induction system	Bing piston valve carburetor

TRANSMISSION
Type	Fichtel & Sachs four-speed, hand operated. Four-disc plate clutch. Reverse gears electrically selected. Noiseless gear shifting by Teleflex cable.
Ratios: 1st	17.70:1
2nd	9.05:1
3rd	6.06:1
4th	4.22:1
Reverse	As above

FINAL DRIVE
Type	Roller chain .5 inch x .25 inch driving rear wheel. Chain in oil bath

CHASSIS & BODY
Frame	Torsion-proof steel tube unitized frame with completely enclosed floor

STEERING
Type	Direct linkage, no steering box
Turns lock to lock	.75
Turning diameter	Approx 29 feet

BRAKES
Type	Internal expanding drum type mechanical brakes activated by cables

SUSPENSION
Type	Each wheel separately suspended with soft rubber torsion springs with hydraulic shock absorbers on all three wheels
Tires	4.40 x 8

ELECTRICAL SYSTEM
Ignition system	Coil, 12 volt (two 6-volt batteries)
Lighting system	90 watt

WEIGHTS AND MEASURES
Wheelbase	79.9 inches
Overall length	111 inches
Overall width	48 inches
Overall height	47.2 inches
Front track width	42.5 inches
Ground clearance	6.5 inches
Curb weight	506 pounds

CAPACITIES
Fuel/oil	3 gallons fuel/oil mixture of which ½-gallon is reserve

PERFORMANCE
Fuel consumption	50-70 mpg
Speed	Approx 62 mph max speed Cruising speed, 45-50 mph

Left: Despite triangular wheel layout, 'schmitt is amazingly stable and flat in cornering. It takes almost suicidal effort to lift a front wheel.

MESSERSCHMITT
continued

possibly one of the finest original examples of the marque in this country, if not the world.

What is a Messerschmitt KR-200 like to drive? It has to be one of the oddest motoring experiences in all the world. One steps into the tiny vehicle as one steps into a bathtub, one leg at a time. First you raise the cockpit all the way. Then you make sure the seat is in the fully retracted position. The front seat rises some six inches up and back on over-center links to facilitate entry. Once the driver is seated, he moves the seat forward, positioning himself comfortably behind the wheel. The top closes quite easily, but it is another matter lifting it again when you want to get out of the vehicle.

Instruments are simple and direct. To the far left is the choke. Directly above the choke, and slightly to the right, is the light switch which controls both fog lights and headlights. There are only two gauges on the instrument panel, the clock to the left of the steering column and the speedometer (measured in kilometers) to the right. To start the vehicle, you must first turn on the petrol. Then turn the ignition key located at the far right of the instrument panel. A red warning light indicates that the generator is on. If you want to go in reverse depress the ignition key before turning it on, and a green light, in addition to the red light, will indicate the reverse sequence. If the engine is cold you will want to use the manually operated choke, but as soon as the engine is running turn the choke back in or over richness may stall the engine.

Getting used to the shifting is something else. A small trigger on the shift lever provides a neutral position between any two gears. Unless you are starting out on an uphill incline you may not want to use first gear at all. Second gear suits denser traffic from

Is Three Enough?

Heinkel, Messerschmitt and Isetta were far from the only three-wheelers. The concept of three-wheeled motorized transportation is as old as the automobile itself, and nearly every industrialized country in the world has produced one or more three-wheelers at one time or another. The most memorable is England's Morgan three-wheeler (see *SIA* #48), and the most fascinating is probably the American dream car, the Davis (see *SIA* #1), of which 17 pilot models were built in the late 1940s.

Like the steam car and the electric, the three-wheeler will always have its exponents, but the practicality of three-wheeled transportation has never washed completely. The primary advantage of the three-wheeler is that it provides semi-conventional transportation for considerably less than the cost of four wheels. Eliminate one wheel and you eliminate the costly drivetrain that goes along with the conventional four-wheeled vehicle. The steering can be direct. The engine can be linked directly to the drive wheel via a chain. The three-wheeler combines the economy of a motor scooter with the passenger and luggage space of a light automobile. Add a third wheel and you can even fully enclose the vehicle. Your end cost is only two or three times that of a motor scooter as opposed to five or six times the cost for a light, four-wheeled passenger car.

There are some sacrifices, however. For a three-wheeler to have the stability of a four-wheeler, close to 75 percent of the weight must be carried on the axle and the tread has to be close to 70 percent of the wheelbase rather than the typical 50 to 55 percent for four-wheeled cars. True, you get increased maneuverability if you put the third wheel in the front. Your turning radius becomes about half that of a normal car. The Davis had a front wheel turning radius of only 13 feet. Streamlining is said to be a real advantage, especially if you put the third wheel in the front as with the Messerschmitt. But just how important is streamlining at city street speeds? Few three-wheelers are built to travel at turnpike speeds, where streamlining is a significant factor. Besides, their economy is basically in their engine size and light weight, not their aerodynamics.

Now what about handling and roadability? Here's where the three-wheelers all fall down. All have poor roll stiffness, that is resistance to body roll when acted upon by centrifugal force in a curve. The third wheel has virtually no roll stiffness at all, therefore the end of the vehicle with the two wheels must do double duty. This end will always tend to steer away from the side force.

In the case of the Davis you get violent oversteer. The rear end is always trying to steer itself out of the turn, so the driver has to steer into the turn to correct. Put the single wheel in back, as with the Messerschmitt, and the characteristic now becomes violent understeer. The front tends to steer out of the turn and the driver has to pull it back in. It is not quite as difficult as with a single wheel in the front, but is tricky, nonetheless. At high speeds it can be downright dangerous.

But weren't you always told the Morgan three-wheeler handled satisfactorily? Actually, it was a very tricky car in corners. Remember, too, the Morgan was a light vehicle with a motorcycle engine. Handling problems greatly compound themselves when you put a modern, high-performance engine of four or more cylinders in a three-wheeled vehicle. The closer you study three-wheelers the more you realize their limitations beyond minimum transportation.

Yet, in this increasingly crowded world they begin to make more and more sense. The three-wheeler could yet emerge as the primary personal transportation of the twenty-first century. Already inflation has put the cost of an automobile out of the reach of millions of city dwellers. Consider, too, that as rail and air transportation continue to improve, there may be less need for the personal highway car with its big engine and attendant size. As energy costs continue to increase and as man continues to cluster in metropolitan areas, the three-wheeler could well become the car of tomorrow.

*Facing page, top: Back seat room is adequate but hardly generous. Below: Gearshift is ratchet type. Trigger on shifter selects neutral. **This page, left:** Side windows slide open for ventilation. **Below:** There's storage space for a handbag behind seat. **Bottom:** Inner lamps are for turn signals.*

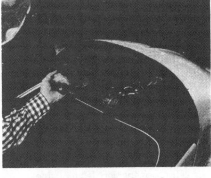

zero to 25 mph. Third is a very flexible gear for flowing traffic in the 18-45 mph range and fourth gear is like an overdrive.

The vehicle travels most efficiently with one adult. A second adult in the tandem rear seat will noticeably affect performance and fuel economy. That passenger will also quickly discover that the sensations for the driver are greatly exaggerated in the rear. A vehicle that is fun for one can be downright scary for the second person. And even 25 mph seems like 50 in the rear seat. With only the driver, the vehicle covers a standing quarter mile in about a half minute, moves along comfortably in the 45-50 mph range and will do 55-60 providing there is no strong headwind.

Driving a Messerschmitt is like riding a bicycle. Only "think" to turn. If you actually force the turn you've probably turned too hard. However, this is an amazingly stable vehicle, and is very difficult to overturn. It gives the sensation of being a lot less stable than it actually is. However, it is possible to lift the outside front wheel in a turn. When you do this, you keep right on turning in the same direction because the vehicle is now, in effect, behaving like a motor scooter. Our driver, Carl Felling, was only able to lift the outside front wheel by shifting his weight to the opposite side and cornering hard at 30 mph, proving how stable this vehicle is. Carl cautions us, however, that such a maneuver should be attempted only by an experienced driver.

Said the British publication *Motor Cycling*, in a 1954 report on the KR-175: "Though the suspension seemed to be somewhat hard, the Messerschmitt handled well, if tending to become skittish on rippled surfaces at speeds of more than 50 mph. Thanks to the direct steering, the driver was always in command of the machine. Any tendency to "break away" on fast corners could be felt immediately and appropriate action taken. Actually, despite its comparative-

ly narrow track and long wheelbase, the "Kabinenroller" was surprisingly stable and could be hauled around corners in a manner which suggested that the makers had incorporated a pot of glue in each wheel." (Bear in mind that the KR-200 had a wider front wheel track and suspension improvements.)

In all, it is an amazing and versatile town car. There are a few drawbacks. In the normal driving position you cannot see your front fenders, and you may, indeed, nearly forget that they are placed well outside of the fuselage. The vehicle is small and low, even by today's small-car standards. While you have a good view of all the traffic around you, not every other driver will see you. Being under 15 horsepower the Kabinenroller is not allowed on freeways in California except in areas where the only road through is a freeway. This is not the kind of a vehicle you could legally drive to a meet in the Los Angeles area if you

lived in San Francisco, unless you wanted to take the scenic route along Highway 101 or take the long haul down old Route 99. But there is one happy factor which differentiates this type of vehicle from nearly every other car of its era. The quarter-century-old Messerschmitt is a vehicle whose time may be yet to come.   □

Acknowledgements and Bibliography
Automobile Quarterly, *Volume 11, Number 2, second quarter 1973; Messerschmitt factory literature and manuals.*
Special thanks to Marilyn and Carl Felling, Topanga, California.

1948 he was testing his first motorized vehicle.

This device, which he called the Flitzer, or "that-which-flits-about," and advertized as *"das Fahrzeug fuer Jedermann"* (Everyman's Car), somewhat resembled the *Kabinenroller* that would come five years later, but it was only a single-seater. The Fend Flitzer had two wheels in front, and a 98cc scooter engine-cum-rear wheel in the back. Its steel frame was covered with pie-section sheets of aluminum molded to form an egg-shaped body that, like a cardboard globe, contained no compound curves but lots of creases. A curiously well-proportioned little contraption with its sporty open cockpit and offset headlight, and very much cheaper than any car available in post-war Germany, the Flitzer seemed to meet the needs of the moment for a couple of hundred local buyers. Fend's small company prospered enough to branch out into development of variations, such as small three-wheelers specialized for the transport of light-duty cargo.

But Herr Fend, criticized in later years for possessing less business acumen than engineering ability and artistic enthusiasm, wanted to grow beyond his resources, and so in 1952 he approached the nearly dormant Messerschmitt factory in nearby Regensburg suggesting a joint venture.

Professor Willy Messerschmitt, something of a maverick himself, had bucked enormous political opposition to contribute the great Bf 109 series as well as the seminal Me 262 twin-engined jet to his country's war effort. Postwar, he was reduced to turning out odd-lot auto parts and sewing machines, and his several factories were running at a fraction of their capacity. His directors were on the lookout for additional product, but they didn't think Fend's oddball Flitzer was it.

Not, that is, until Prof. Willy himself was introduced into the discussions and immediately recognized Fritz's as the friendly face of a technical officer who supported him during his struggle to "sell" the 262 jet to the government. The famous designer told his board, yes, this baby auto *is* something Messerschmitt wanted to take on. On just such chance human encounters are the formal fortunes of history written.

The *Kabinenroller* thus had Messerschmitt's name on it, but it remained essentially Fritz Fend's project, and to the end of the line a decade later, he was personally involved in design and development. Indeed, Fend even performed the test riding for the *Kabinenroller*, ran the little three-wheelers in races (!), and was one of the intrepid pilots of a special *Kabinenroller* streamliner for a 1955 record run on the Hockenheim road circuit. (Photos exist of the streamliner also running on the high brick bankings of the Avus track outside Berlin.) This adventure was a great success, with the streamliner running to a peak velocity of a screaming, terrifying, mind-boggling 74 mph, resulting in 22 new speed records over a 24-hour period.

All of this was possible because the two-seat *Kabinenroller* caught on—as with the Flitzer, but moreso, it filled a need, and the little trike its fans nicknamed the "Karo" could be seen by the thousands all over Germany during the Fifties and well into the Sixties.

The Messerschmitt you see here is one of the very first. Indeed, it may be the oldest existing example—its owner knows of no serial number lower than his KR175's Fahrgestell Nr. (chassis number) of 1351.

Yes, he does plan to restore it. Although it may be a while before he gets to it. Paul Prince, a designer of exhibitions for the University of California at Santa Barbara, would have gotten along with Fritz Fend. A man who neatly blends the eye of an artist with the hands of a mechanic, Prince is inevitably awash in such projects. To keep from being overwhelmed, he's trying to grasp the neck of just one restoration at a time. First comes the completion of his flathead Ford hot rod, "just like the one I had when I was a kid." Then he wants to get his aged Alfa Romeo coupe fixed up, "so I have something to drive." Only then, he vows, will he allow himself to mess with his Messerschmitt. (However, from the glint that came into his artistic eye as he was showing us all the oddities of his battered, but intact, old Karo, his enthusiasm may have been rekindled enough to reorder his job-list. . . .)

Prince's KR175 is an early enough model that it appears primitive even by *Kabinenroller* standards. For steering, it has something very like a handlebar—later versions were equipped with a slightly more car-like plastic yoke. However, this particular vehicle was made late enough to have a car-type clutch pedal, rather than the original lever mounted to the gearshift.

With a central seat under its bubble canopy, the *Kabinenroller* naturally makes a person think: Fighter! The big plastic dome is even hinged at the right, as was the Bf 109's canopy. To get aboard, you step over the left-side bodysill, support yourself with your hands, and wiggle your legs down as

KABINENROLLER

In the late 1930s, Willy Messerschmitt's sensational new aircraft—later to become the backbone of the Luftwaffe as the Bf 109—set speed records unsurpassed until the 1960s. The Kabinenroller was a good deal slower.

Photo: Motor-Presse-Archiv-Stuttgart

Photo: Albin Joseph

you would into a pursuit plane's cockpit. While entering it's easier, but not essential, to make use of a pivoting linkage that first moves the driver's seat up and back. An adequate 23.5 inches of clearance is available between sidewalls at the upper-arm level, and the Messerschmitt provides plenty of leg room. Even with the canopy shut you still don't really feel enclosed. You can see in every direction almost as well as . . . from a ''real'' Messerschmitt? Better. Willy's war-machine was not noted for its outstanding visibility: long wing-blades

would have been knifing out under your shoulders, a great masculine engine-snout with warty gun-covers jutting out ahead. But seated normally in the KR175, you have almost nothing of it in view—indeed, to see the edges of the front wheel fairings adjacent to your ankles you have to crane your neck. The only external portion of the vehicle always in your vision is the pair of small headlight fairings, which generates an image of gentle femininity if your mind turns to such things.

But the Karo is not much at all like a

plane, car or motorcycle to operate. Growing out of the floor are normal automotive brake and clutch pedals, but a twist-grip on the handlebar controls the throttle. A third foot pedal, on the far left, links to the engine's kick-starter mechanism. (This early example has only a six-volt electrical system; later KRs got 12 volts and electric starters.) By your right knee is a parking brake lever. A little closer, thigh-level, sits the shift-lever, which operates the transmission motorcycle-style—i.e., a series of pushes or pulls cycles the cog box up or down through four forward ratios. A neutral finder is present in the form of a handlebar-mounted lever. The KR175 has no reverse—you either climb out and push, or carry a stick. (The successor model, the KR200, provided reversing capability by the simple expedient of running its two-stroke engine backwards!)

Because the handlebar is linked to the tiny front wheels as directly as in a go-kart or four-wheel ATV, with no reduction gearing, Messerschmitt steering is decidedly abrupt. The shifter's location on the right and the twistgrip's on the left gives a contemporary motorcyclist something extra to get used to—and the throttle twists ''backwards.''

The simple dash is dominated by a radio, dead center (missing in the photos), and by a 100-kilometer speedometer with a clock to right and left, respectively. The ignition switch is far left, the light switch far right. And that's all there is to running a *Kabinenroller*. Oh, except for the windshield wiper, which you operate manually through a trigger mechanism at the lower right corner of the windshield. The shield itself is, of course, a glass insert.

Similarly, the KR175 is structurally simple. A steel tube runs along either side, slanting up from front to rear, and from these tubes hangs the pressed-steel floorpan. The external bodywork is also pressed steel. Front-wheel suspension is by means of rubber blocks, with no particular attention paid to preserving geometry. At the rear, the entire engine/wheel package pivots on springs at the top front. At the bottom front one would find more rubber on the original vehicle—four steel coils in tension are fitted to the Prince vehicle in its present condition. The corners of this tricycle are innocent of shock absorbers. All three mount the same size tire, 4.00 by 8.0 inches. Wheelbase is 80 inches, front track 36.25, and we measured the maximum body width at 48.75. We did not bother to measure the empty weight of this presently incomplete vehicle, but figures between 407 and 485 pounds appear in the literature. *Zulassig Gesamtgewicht*, or gross weight, as stated on the chassis plate is 360 kilos, or 794 pounds, so your passengers should be petite and your duffel discreet.

KABINENROLLER

Who owns the road? Inside the cockpit canopy, the Messerschmitt pilot confronts ''direct'' controls, tiny spaces, anthropomorphic bodywork.

KABINENROLLER

The Messerschmitt is a period piece: crude, unorthodox, out of scale, yet its homely, cocoon shape captures undeniable charm.

The fan-cooled Sachs single, with a bore of 62mm and a stroke of 58mm, has a compression ratio of 6.6:1 and runs on three gallons of 25:1 gas/oil mix. At 5250 rpm the engine puts out an alleged 9.5 maximum horsepower through a three-disc, cork-lined clutch and a 1/2 x 5/16 chain to the rear tire.

According to Paul Prince, driving the early-'50s Messerschmitt on the 1980s streets is a gas. "For one thing, everybody's looking at you. They can't figure out what it is.

"It takes quite a while to get acclimated. The steering is right now, and you over-control a lot at first. It feels odd to have the front wheels under your feet, and turning makes your body move oddly in space. But it's like surfing: You start out wobbly, but after a while you know what you're doing and gain confidence." He adds that, even with the sliding side windows out, the sun's greenhouse effect makes under-bubble temperatures uncomfortable.

We gained further insight, as well as helpful research data, from a Messerschmitt man in Santa Rosa, California. Gary Koehler says that with only the driver aboard, the *Kabinenroller* handles quite well. "With that one-to-one ratio, you don't do much steering. It's more like a Kart in the way it steers. The Messerschmitt 'whips.' It's zippy, and very controllable. You can zig and zag between marker cones on the road real well! And it's easy to get up on two wheels, like you're sailing. That's fun.

"But with a passenger on board, it gets pretty scary. The rear end sways on that little tire. The poor sucker in back is biting his nails."

Koehler added that the bubblecar was easy to work on, which was good, because an owner had to de-coke the head every once in a while. Rust in the bellypan was a widespread longevity problem. He also recalled for us the simple, direct heating system: "A hose that ducted engine air forward, and you had a choice of aiming it up at the windshield, or down at your crotch."

In time, the *Kabinenroller* evolved from the original 175 model to the much-improved 200cc model of 1955, and on to a 500cc, four-wheeled "Tiger" version in 1958. By that time, Fritz Fend was back on his own. As of 1956, Messerschmitt was allowed back into aircraft production, and cabin-scooters had to go. They went with the professor's blessings and permission for continued use of his name, but from then on "Messerschmitts" were made by a new firm, *Farhzeug und Maschinenbau, Regensburg,* or FMR.

For a while, Karo popularity continued at levels high enough to support several models, including a true convertible. But the pace of West Germany's postwar "economic wonder" was quite rapidly outstripping the minimal vehicle idea with more and more people able to afford real cars. This time Fend, so ready with the right product for a decade earlier, did not pick up on the trend. He concentrated development on details, not the concept as a whole. Meanwhile, one of the manufacturers that had followed Fend's footsteps to build a tiny bubblecar of its own, BMW with its Isetta, turned back to automobiles; at the same time, BMW aggressively pursued the slowly growing sports-motorcycle market. Once again, the historic schism between automobiles and bikes deepened.

If there had once existed a chance to truly blend the best features of two vehicular worlds, the moment was gone. The status-conscious German public began to see the *Kabinenroller* as one of history's curious footnotes, almost a minor automotive joke, a quirky vehicle for a quirky and difficult time now past. The last "Messerschmitt" left the assembly line in 1962. ∎

MESSERSCHMITT

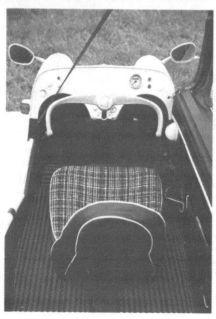

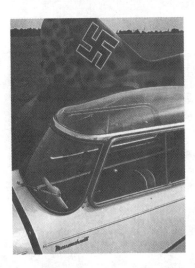

MYTHS

N ot many months ago a colleague mentioned the old story that after the Second World War the German Messerschmitt aircraft factory had so many fighter cockpits left over that the company decided to make good use of them by designing a bubble car round them.

I had heard a similar tale many years ago and was quite happy to go along with his theory. After all, you could imagine an Me109 crashing and losing its wings and fuselage, leaving intact just the cockpit area. Flatten the nose, add three wheels, stick a two-stroke single up its tail, and there you have a Messerschmitt for groundwork only.

Last year at the Messerschmitt Owners' Club rally I learned that this piece of motoring folklore was complete nonsense, romantic or otherwise. I could tell from the tones of the replies that members were weary of dispelling this popular myth.

Direst enemy

Surely, though, there must be some connection between the Supermarine Spitfire's direst enemy and the tripod tandem bearing the Messerschmitt monicker?

Indeed there is. The car was built at Regensburg, West Germany, in the plant where the formidable Me109 took shape. Regensburg also spawned the Me163 rocket fighter, and the Me262 twin-engined jet fighter which outclassed everything the RAF had at its disposal.

Production of the first Messerschmitt three-wheeler actually commenced in 1953, eight years after war had ended in Europe.

The 175cc KR175 made its debut at the Geneva Motor Show, Switzerland. It was another alternative for those seeking inexpensive personal transport in the comparatively austere years of the immediate post-war period.

The KR175 arrived on British soil looking for friends a year later. It came minus suspension, reverse gear and heater. But it had cheeky looks and a certain charm.

And apparently the Messerschmitt did make friends, becoming quite fashionable in cities and major towns throughout Europe, particularly after it was upgraded in 1955 to become the KR200 with suspension and a bigger 191cc Sachs engine giving — wait for it — 9.7bhp at 5250rpm.

Frontal area

According to the owner's handbook unladen weight is only 510lb. While the resulting power-to-weight ratio is hardly earth-shattering you can deduce that the four-speed motor could give the Kabinenroller (cabin scooter) a reasonable turn of speed.

Only four feet high, 4ft 2in in overall width, and 9ft 4in in overall length, its minimal frontal area assists speed and frugal fuel consumption.

Colin Archer from Woking, Surrey, who owns the first KR200 Cabriolet imported to Britain, says he averages 60mpg, "and it's very comfortable and enjoyable to drive."

When *Motor Cycling with Scooter Weekly* tested the KR200 (February 26, 1959) it was reported that on a level road the test model would maintain a true 56mph on full throttle.

Scooter and Three Wheeler tested the same example (April 1959) and reported that "sedate cruising at a steady 40mph on country roads proved most enjoyable, particularly when a petroil (24:1) consumption figure of 87.2mpg was recorded."

A tuned version of the KR200 had gained credit for the marque in 1955 by establishing 25 class long-distance and endurance world

CONTINUED ON PAGE 169

Heard the one about Messerschmitt bubble cars being built from redundant fighter aircraft cockpits? Sadly, it is nothing more than motoring folklore but Brian Crichton found that the cars have a fascinating history

Entry to the cockpit is through the side-hinged canopy (above), making the car feel even more aircraft-like. The two-stroke engine is accessible through the hinged tail section (below)

BACK TO BACK

LOW *fliers*

From the makers of caravans and aeroplanes came two of the smallest sports cars of the fifties, here driven by Stephen Bayley.
Pictures: Andrew Yeadon

I drive an Audi Quattro and scarcely leave the London postal districts which the nuts and berries folk never reach. The car and its peculiar urban environment are all products of recent technical and economic miracles. Arriving at Herne Hill cycle stadium to drive two products of an earlier *Wirtschaftswunder* was a bit like one of those stock scenes from old science fiction movies where the hero leaves his spaceship to discover – blinking – a lost world. Here was an alien landscape and... classic cars. Or, at least, classic tiny cars. I learnt from my first close encounter with owners that they like to talk of oomph and bumph. Do you want to tell me something about the car? I said to Paul Fitness, looking in polite wonder at a 1960 Berkeley B-105.

"I suppose I've had it about 12 months now. I've done it up *in order to use it.*"

I find that the distribution of Berkeleys is very uneven. People seem – like me – either to never even have seen one before, or actually to own rather a lot:

"This is about my fourth and I've got another two at home as well."

What are they made of?

"One is the three-cylinder two-stroke waiting to be reassembled, and a twin-cylinder three-wheeler also waiting to be reassembled. I'm not really interested in prizes. I can go out in the rain and I can drive through floods and I don't mind. It's really great. It was built for a top speed of 105mph. Even now I have great fun with a lot of modern machinery. This green one's got a Royal Enfield 692cc Constellation. They also did a slightly detuned version called the Super Meteor."

Built like a gun, goes like a bullet, I'm told they used to say. I've also heard them called Royal Oilfield.

"Yeah. This one's got high-lift cams and different pistons, which gives it that extra bit of oomph. Coming down the M25 I like to cruise at 50-55mph. I did read that these Berkeleys were the first vehicles to achieve 100 per cent braking ability. Mind you, they *were* very light."

Certainly looks it. Mind you, I don't yet even know which wheels are driven.

"There's the primary chain which drives an Albion multi-plate clutch and the differential is of Berkeley's own design. It was originally oil-filled, but as the planet gears are rotating in the aluminium casing,

they open up the holes in the sides so they lose all the oil. The way around that is to fill it with grease. Never had any problems with *that*."

But do the Audi engineers they show at the film club know this? What about the electronics, the engine management system?

"Well, it's a c-o-n-v-e-n-t-i-o-n-a-l starter, in fact a Triumph 1300 reverse throw inside a Morris Minor case. They used to race these with very good results. Still hold track records here and there."

But not at Herne Hill, at least not in my hands. The moment when I have to DRIVE it is approaching. I put off the evil hour, observing that the engine looks huge in a tiny car that is only ten foot five and a half inches long and made out of old suitcases, but console myself by allowing that the specification sounds sophisticated: aluminium alloy engine, plastic monocoque with aluminium and steel gussets at stress points, wheels with detachable rims bolted directly to brake drums, independent suspension with unequal arms at the front, swing axles at the rear and coil springs all round and – I discover at last – front-wheel drive.

From the front, the car looks a bit of a pig. Its predecessor had a petite charm which Honda later captured in its first miniature sports cars, but the requirement to accommodate the perpendicular built-like-a-gun-goes-like-a-bullet has caused ugly elephantiasis of the snout, an effect made worse by the legal height of the lights. But here we go. I am not terribly big and I am quite agile, but the Berkeley cockpit looks very small and you have to go through an awkward contortion to get into it. Once settled in, it is surprisingly roomy and the driving position is not as Godawful as you feared, although the wheel is close to the chest by contemporary standards. It is more similar in character to a cheap caravan than an expensive Swiss watch, which is to be expected, as cheap caravans were Berkeley's business until in 1956 they diversified into miniature sports cars when they bought Laurie Bond's designs for this thing I have to drive at the Herne Hill cycle track.

The engine starts as though there's something terribly wrong with it, but as it doesn't stop you have to assume everything's in order. All the controls come to hand and foot well enough. I can't remember the clutch so it must have been fine, but the quadrant gearshift (like a Harley-Davidson) is unforgettable and I'm astonished I can get it to work. I think I probably went straight from first to third or maybe even to fifth, but then I have no idea about how many gears it actually had. Nor do I know how quickly I was going, but it seemed like at least half a Quattro, even in first. The noise was historical and not displeasing, the brakes very heavy for such a titchy machine, but encouragingly effective. Even I could tell that this motorcyclised tub was only likely to be

Top: The Berkeley in brand-new, standard trim – note afterthought over-riders. Above: Light weight and high power made the B95 and B105 ideal budget racing cars – this is J.I. Goddard-Watts' car, with extra, home-made, grilles for brake cooling. Below: Entertaining handling – this is George Catt at AMOC Silverstone, August 1960, having fun

stable as long as the power was on.

Rather like a Zeppelin, in fact. This observation brings us directly to the Messerschmitt, another little sports car arising out of identical historical conditions, but with substantially different results. The poor old Berkeley is a product of the awfulness of Britain in the immediate post-war years, the world of gristle rissoles, flour soup, caravans and tawdry, talentless Shepperton film stars. On the other hand, Professor Willi Messerschmitt's work on gliders was inspired by his experience of airships which gave him an almost obsessive interest in light weight and low frontal area. Like every other German engineer, *"Weniger Teile!"* (Fewer parts!) was his unspoken motto. What the caravan folk would not have understood was that each of the *Teile* had to be perfectly formed.

The Germans are an ingenious lot and readers of Reinhard Lintelmann's *Deutsche Roller und Kleinwagen der Funfziger Jahre* (1986) will boggle at the variety, creativity and quality of post-war German miniature cars. The Messerschmitt KR200, designed by Fritz Fend, appeared in 1955. Owners are the bumph rather than the oomph lot. Mr Barry Rossiter explained:

"I've had the car (the extraordinary four-wheeled TG500) about four years. It's a 500cc twin, designed by Sachs. It needed rebuilding. I've come up from Somerset. One hundred and twenty miles. It's very tail-happy. The gear-change is... different. I've only ever really been into Schmitts."

With such devotion to the Karo (Kabinenroller) you feel reluctant to expose something so precious to risk, but curiosity is more powerful than decency, so in I get. The driver's seat has a fascinating over-centre action to allow access to the intimate, not to

Clockwise from below, second from left: Tandem seating is straight from a fighter aircraft; Sachs 500cc twin makes the Tiger "very tail-happy"; plastic canopy raises manually to admit occupants; owner Barry Rossiter – "I've only ever really been into 'Schmitts"; the TG500 – 'a snugly-engineered miniature Volkswagen', says Bayley; flop-top is detachable

Clockwise from above, second from right: 'Royal Oilfield' 692cc engine – built like a gun, goes like a bullet; it's a pretty car from this angle, perhaps with slight Ferrari overtones; B105 owner Paul Fitness – "this is my fourth"; the Berkeley was built to do 105mph; 'from the front the car looks a bit of a pig' (Bayley)

say cramped, cockpit. Sitting under a perspex dome is a weird, but exciting experience. It is like riding a bike with a fixed clear shroud. You can see why they called the first Schmitt *'Der duserjager des kleinen Mannes'* (the ordinary man's jet) and why the brochures advertised its *'monoposto Effekt'*, a curiously leaden feeling possibly provided by the fact that you're sitting right over the centre of gravity.

Perhaps strangely for a tiny machine with no weight over the front wheels, the tiller steering is very heavy (although contemporary reports always described it as light). The engine spins well and the car seems fast. It hunkers into the banked corners nicely, more nicely still if you are in the appropriate gear. I could only find them in entirely random access so I had to take instruction from Barry in the back seat...

Covering myself, I venture that I remember Dr Porsche writing that his cars provided 'driving in its purest form'. I'm not sure I'd say that, but it's certainly a unique experience...

"Well, you've certainly got the knack with the gearchange."

There is a noise like a kitchen machine with a stripped crown wheel and pinion. Tell me again what the technique is for the up-changes.

"Well, you need to get some revs on. Up to four or four and a half. You want a s-l-o-w change, not a fast one. Then you just push the clutch in and the gearlever at the same time. Five and a half was reckoned to be top revs."

Above: C.N. Morrell in his TG500 has an anxious moment during the 1964 Land's End Trial after the hump at Bluehills Mine. Above: BOR 500 was a regular campaigner in the hands of owner Ken Piper in the early sixties, taking part in trials and rallies – note the lack of canopy. Below: Hard cornering produces some interesting suspension movements

The Berkeley is perhaps the better car, but the Messerschmitt is the better machine

Fantastic. Oh look, my son's carrying back the rugola as well. It's a fantastic car and you've made me want one. What does one of these cost now?

"It's very difficult. About ten thousand."

A student of the old *Hochschule für Gestaltung in Ulm*, Germany's outstanding post-war design school, was recently speaking at a London conference about the atmosphere of Germany in the fifties. He said: "One can't talk about *design* in the years immediately following 1945: the problem quite simply was how to make a usable cooking pot out of a steel helmet."

Or a sporting *Kleinwagen* or cheap plastic sportscar with a motorcycle engine. Both the Berkeley and the Messerschmitt are four-wheel bikes, with many of the compromises and advantages that suggests. They offer a vivid portent of why, 30 years later, only the Federal Republic is building cars like the Audi Quattro.

One is a snugly engineered miniature Volkswagen, purpose-built, blower-cooled, geared differential and ingenious interior details. The other is an amusing, but frankly shabby bitsa with a brutish lump of old (1952) engine bracket-mounted on to a drive mechanism which would have embarrassed Archimedes.

Of course, while the little Messerschmitt opened up a path that headed to nowhere, it emits a feeling of industrial confidence and competence. Yet the Berkeley has ideas that Colin Chapman later exploited to better effect.

You feel that with better management and better quality control, the Berkeley might have made it. It is perhaps the better car, but the Messerschmitt is the better machine.

The simple Sprite and the fabulous Mini drove the Berkeley out of business. In Germany, increasing wealth stimulated a demand for larger cars which put Messerschmitt out of the market. The Panter family who owned Berkeley went the way of the Beverly sisters. Professor Messerschmitt went on to found MBB, Europe's third largest aerospace concern. And it shows.

Small change

PHOTOS ALLAN LEVY

Aircraft engineer Roger Blaikie used ingenious means to turn an incomplete Messerschmitt microcar into a show winner. He tells Nick Larkin how he did it

When Roger Blaikie proudly opens his lock-up garage to unveil the car ne spent 18 months restoring, you get the awful feeling that thieves have been there first.

It's only after anxiously peering into the gloom that you discover a vehicle *is* there. A minute, bright red, Messerschmitt three-wheeler sits in the middle of the floor, like a friendly pet waiting for walkies, and dwarfed by the breeze block walls around it.

Roger reveals that the car is a mere 9ft 2in long, and 4ft high, weighing 510lb unladen. The 8in wheels give a six-inch ground clearance. There are no doors, merely an aeroplane-style cockpit which lifts up to let you in and out. Power comes from a 191cc rear-mounted Sachs single-cylinder engine.

The car may be no behemoth, but as Roger testifies, a lot of restoration trouble can come wrapped in small packages. He's had to overcome greater hurdles than those faced by many rebuilders of cars four times the 'schmitt's size.

An aircraft fitting instructor at British Aerospace in Hatfield, Herts, Roger, 37, has been restoring cars and motorcycles since he was 14, his previous projects including a Rover 2000, Triumph GT6, a Bedford caravanette and numerous motorcycles.

Roger had been looking for a Messerschmitt for two reasons: 'I was offered one for £10 20 years ago and turned it down, regretting my decision ever since,' he recalls. 'Also, my girlfriend

Roger lowers the cockpit ready for another run

Indicators were among many hard-to-find components

Ann saw one at a show and decided a 'schmitt was definitely the car I should restore next.'

After a lot of fruitless searching, Roger advertised in *Exchange and Mart* for a suitable car. On grabbing the magazine to see if his ad had appeared, he noticed a Messershmitt for sale, and tore over to Milton Keynes to snap it up.

'It must have been the worst example anywhere. The floor reminded me of a

rotting leaf — it was totally porous with huge holes in. All the the body panels were well gone,' recalled Roger.

The list of horrors goes on. There wasn't an engine, though the car came with a piston, half a crankcase, two cylinder heads and a carburettor casting.

There were no lights, the wiring was lethal, and the entire interior beyond repair.

'I know it was bad, and even some people in the owners club said I must be

mad to attempt a restoration. But I knew I could see it through,' Roger said.

He got to work immediately. The first stage of this complex restoration involved commandeering *cornflakes packets* from all and sundry, not to coat with underseal as chassis repair sections, but to act as templates for new body panels.

He stretched the cardboard over what remained of the original sections, and marked the position of bends. Then, using flat sheets of 20 gauge steel, he recreated the bends using a folder, and formed the curves using panel beating equipment.

The side panels were MIG-welded to the car's tubular monocoque, and a lip was made at the bottom of each to support the floor.

A new floor was ingeniously made by constructing a mould, using hardwood to match the original ribbed floor panel and mounting it on an aluminium backplate. This was then taken to an engineering company, and

the pattern stamped out on an industrial press.

'I reckoned the wood would hold just enough for one application. Fortunately I was right,' said Roger. But he added: 'Probably the most regrettable thing about the whole restoration was that I didn't join the Messerschmitt Owners Club until towards the end. Had I done so, I'd have found out I could have got a floor from them!'

The bottom six inches of the car's 'nose' was completely rotten, and new ones are unavailable, so Roger cut out the old metal and welded in new. Two new inner front wheelarches were shaped to fit.

With the bodywork complete, two coats of filler primer were applied, followed by several top coats. 'I can't remember exactly how many. I just kept on going.' Each coat was flatted by hand, all the work taking place in Roger's single garage.

He chose a current Rover shade of bright red as it was

Six inches of new metal were needed to repair frontal corrosion

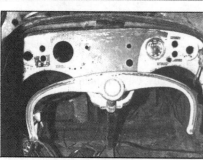

Side panels were remodelled using cereal packets as templates

Interior was a sorry mixture of decay and missing parts

Parts from several countries were needed to complete the 191cc engine

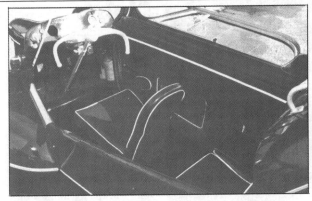
Car was completely reupholstered to original specification

Excellent accessibility to 'schmitt's miniscule engine bay

close to the original colour, and stonechips could easily be touched in.

A complete set of seat and interior panels had to be made up, mainly by using the originals as patterns. The steel and aluminum canopy was repaired by welding. The front windscreen was intact, but new sidescreens were made by cutting double glazing acrylic with a hacksaw. A replacement wiper motor is believed to have originally adorned a milk float!

A friend, Geoff Marshall, made up a new wiring loom, and a new speedometer was fitted. Front light surrounds were made up on a lathe.

The cable brakes were rebushed throughout, and new shoes came from the owners club. New bushes were needed for the rear suspension, a strange swing arm device made mainly of rubber, which had survived surprisingly intact.

Now came the biggest problem — the incomplete engine. 'The car took me a year, working practically every night to get rolling, and another six months to get running properly,' said Roger.

He began advertising extensively for parts, and tracking them down both in Britain and abroad. 'I felt I'd travelled the earth for that engine by the time I'd finished.'

A good piston came from the owners club. He managed to get pieces of crankshaft, and ended up putting two together, making up his own bearings. Neither led to the engine running perfectly, but luckily Roger managed to get a complete item from Germany, for £130.

'Putting that engine together was very much a trial and error affair. Luckily I managed to get hold of a workshop manual, and even more luckily it had been translated from German,' said Roger.

Eventually, the 'schmitt was running well, and, after the testers had got over the shock of what was before them, sailed through its MoT.

The 1960 car picked up an award for the best restoration from a wreck at the Messerschmitt Owners Cub annual rally earlier this year,

Messerschmitt specification

Engine	191cc single-cylinder two-stroke
Top speed	60mph approx
0-60mph	hopefully
Max bhp	9.7 @ 5250rpm
Mpg	60
Compression	6.3:1

but Roger certainly hasn't turned the car into a cosseted concours toy. In fact as soon as the 'schmitt was legally roadworthy he road tested it — by driving 892 miles to Sweden and back. He raised £1000 in sponsorship for a local school for the handicapped by taking the car to the Swedish Microcar Rally at Trollhatten. 'It went like a dream. The only problem was a return spring on the clutch coming off.'

A quick spurt up the road is all that's needed to fully understand Roger's mania for microcars.

After lifting the canopy, access to the driver's seat is easy, as it slides upwards and backwards to let you in, reversing the procedure when you're installed. 'It's this sort of thing which

Schmitt was resprayed in this cramped lock-up garage

makes you admire ingenious German engineering,' says Roger.

The petrol is switched on by a lever on the rear firewall, and a quick turn of the ignition key persuades the 191cc of raw power behind you into life. The passenger sits, legs askew behind the driver.

The car has four forward and four reverse gears, meaning it apparently can go backwards as quickly as it will go forwards, a feat not to be tested on the Queen's highway by an inexperienced microcar pilot.

The gearlever itself sticks out at your right, a little bit like one of those old-fashioned domestic water pumps. Changing gear is accomplished by merely moving the lever backwards and forwards, again like ye olde pump, and a system you're convinced won't work till you try it.

The handbrake on your left looks like one of the legs ripped off once of those awful wire sixties fruit bowls.

The clutch is positive enough, and the car accelerates sufficiently quickly not to be an embarrassment in modern traffic. In fact it will, in Roger's hands, bowl along at 60mph.

Steering is accomplished by what looks like a tiller in front of you, by which you can only turn the wheels when the car's in motion.

Ride is understandably on the hard side, but the 'schmitt is fairly comfortable for short distances, and, of course, you're travelling cabriolet-style. Hard cornering would take some practice, but gentle bends are

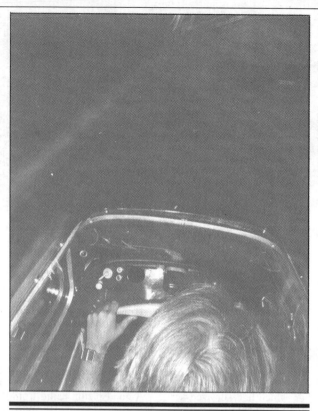

The engine makes a noise similar to a dog barking when the car slows down

tackled with ease

Once on the move, it appears that whole streets stop to watch the bizarre creation in motion. Schoolgirls giggle, children point, pensioners and building site labourers stare transfixed.

The engine makes a noise similar to a dog barking when the car slows down, giving further credence to the theory that a 'schmitt is,

in fact, alive.

Even the 'frogeye' Sprite can't match the Messerschmitt's facial expression. This car, produced by the man who designed the legendary Second World War fighter planes, has almost an accusing look, so much so that you wouldn't, for example, pursue any courting activities where it could see you (you

Clubs

Messerschmitt Owners Club, Birches, Ashmore Lane, Rusper, Horsham, West Sussex RH12 4PS. (0293 871417)

Messerschmitt Enthusiasts Club, 5 The Green, Highworth, Swindon, Wilts SN6 7DB

certainly couldn't get up to such mischief *inside* the car, anyway), or, with the car in your workshop, dunk a biscuit during your tea break.

Roger says he spent around £1000 restoring the 'schmitt, though having an engineering background helped keep the cost down.

He praises the owners club for its help and advice, and reckons girlfiend Anne, who not only faced the 'schmitt ride to Sweden unflinchingly, but found herself riding pillion on the back of a previous project — a Honda Goldwing motorcycle — to Greece. 'Anne's great,. If all other halves had her attitude, there'd be a lot more cars restored.'

Enthusiasts will now pay £4000 for a good Messerschmitt, but Roger, who is about to begin work on two Berkeleys, merely wants to enjoy the little red devil to the full.

'I have had people say I shouldn't drive a car like this regularly,' says Roger, who also drives a 1981 BMW 316. 'But my logic is that the 'schmitt is an interesting car, which I didn't build as an investment and don't want to think of as such. It's a far more rewarding experiece just driving it!' ∎

Schmitt willingly goes places where no car has ever been seen before

Roger Blaikie is still looking for an original clock and petrol cap for the car, along with anything else Messerschmitt. Write to him via *Popular Classics,* and we'll forward the letters.

A melange of Messerschmitts at the 1955 Swiss Motorcycle Show (yes, the Swiss Motorcycle Show!)

BUBBLE & SQUEAK

The Messerschmitt was designed to put post war Germany back on the road, *Mick Walker* looks at the history of this fraught factory.

Like Heinkel, Messerschmitt was already a name internationally known for the company's aviation exploits. In fact, together with the British Spitfire, the Messerschmitt 109 is arguably the best-known aircraft of all time. Certainly, more 109s were manufactured during World War II than any other aircraft.

In many ways it is a miracle that the firm survived long enough to produce any of them, for during the 1920s they had encountered a series of financial problems which saw them virtually bankrupt at least twice.

Messerschmitt's birth can be traced back to October 1922, when the Udet Flugzeugbau - headed by the leading German pilot Ernst Udet - was established with the primary aim of building light sports and training aeroplanes. But in July 1926 another company called Bayerische Flugzeug (BFW) was created by the Bavarian State Government and a banking house, and this ultimately took over Udet Flugzeugbau.

The Messerschmitt name first appeared

in yet another Bavarian aviation company headed by Dipl Ing Willy Messerschmitt, a graduate of the Munich Technical High School who had designed and built several successful gliders and sailplanes during the early '20s. Messerschmitt had formed his own company in 1923, and in September 1927, Messerschmitt-Flugzeugbau GmbH joined forces with BFW and embarked on an extensive development programme - to extensive as events were to prove, with the result that June 1931 saw the company file for bankruptcy.

However, Messerschmitt himself had managed somehow to retain the design and patent rights for his creations, and with a new loan was eventually able to resume business during early 1933. After winning a contract from the Luftwaffe (German Air Force) to build their new fighter monoplane in 1934, the company never looked back until the end of World War II over a decade later - by which time the name Messerschmitt was known the world over.

At war's end in May 1945, the story now shifts to an engineer named Fritz Fend

who had worked on aircraft design during the war, and had come into contact with Messerschmitt at that time. With the post-war ban on munitions work, Fend had set up his own engineering consultancy and like Farinelli in Italy (the man who started Ducati off on two wheels), had already visualised the transport problems facing a country shattered by six years of war. However, his design was not for two wheels, but three - and in the very month that the European sector of the war came to an end his first design was conceived. At first he worked on invalid carriages designed for hand propulsion but by the summer of 1946 the first prototype of a single seater machine powered by a 50cc two-stroke was on the road, and by 1949 a company had been formed to manufacture the new machine.

The Fend concern was based in Rosenheim, Bavaria, where they built a range of small three-wheeled vehicles called the Fend Flitzer powered first by 98cc Fichtel and Sachs engines, later by Riedel engines. The Flitzer could accommodate two people and was the earliest form of what was later to be called

This was just about the last of the 41,190 trikes to see production, and it appeared in June 1964

the bubble car, predating all the other designs which appeared throughout Germany in the 1950s. Fend also produced various commercial three-wheelers with the driver (rider?) seated at the back on what was essentially the rear half of a scooter fixed behind a goods carrying chassis - the reverse of the system used in Italy.

In 1952, after a chance meeting with Messerschmitt in 1951, Fend joined forces with the much larger Messerschmitt concern who were restricted from aviation involvement by the terms of the armistice, and had seen the potential offered by at least one of Fend's designs. The Fend-Kabinenroller (cabin scooter) was a new design constructed at the Regensburg aircraft works and was first seen at the Geneva Motor Show in Switzerland in March 1953 where it created something of a furore. Quite simply, nothing had ever been seen like it before, - a sleek two-seat aircraft with no wings, tailplane or propeller!

As one of the attending press corps enthusiastically described it, the 'latest effort from the Messerschmitt aircraft factory was an attractive three wheeler saloon in which the two bucket seats are arranged in tandem'. And indeed, its construction is worth looking at in detail as although many later 'bubble cars' appeared, none captured the public imagination in quite the same way as the unique Fend-inspired design.

The body construction was a steel tubular frame panelled in sheet steel with two seats and a set of handlebars for the driver. Its overall dimensions were just 48in high, 50in wide and 112in long. Steering was direct and the turning circle was just over 27ft. A pair of headlights were faired into the nose and red light trafficators were provided on the front wings and rear of the body.

Everything above the occupants' chest level was transparent, with the sides, domed top and back in plastic, and a windscreen in safety glass with a single wiper blade which was cable-operated from a trigger on the handlebar. Sliding windows were provided on one side, with a

corresponding removable panel on the other, and the whole saloon top was hinged along the right side in a similar manner to many British family sidecars of the period. When the top was hinged back for entry of the driver and passenger, a panel on the left was raised with it which opened the vehicle down to driving seat level to make entry and exit much easier. There was space for luggage behind the passenger seat which could be removed if more load-carrying capacity was needed.

The prototype was initially powered by a 150cc Fichtel and Sachs two-stroke single, but for the production models this was replaced by a 174cc unit (62 x 58mm) with a cast-iron barrel, alloy head with 6.6 to 1 compression ratio, and battery and coil ignition. There was a 12 Volt combined dynamo and starter made by Siba and a cooling fan which formed the engine flywheel. A four-speed gearbox (with no reverse) in unit with the engine was used, and this had a positive-stop gearchange mechanism operated by a forward-and-back hand lever positioned on the right of the driving seat. This lever operated the clutch automatically when a gear change was made, although later models had conventional car-type clutch pedal. Drive was to the rear wheel, and all

The 1959 Sport Cabriolet KR200 was introduced to the public in October of the previous year

three wheels carried 4.00 x 8in tyres and coupled brakes which were cable-operated by a foot pedal.

The machine's 9hp power output coupled with low weight (155kg) and small frontal area gave it a speed of 50mph, which was comparable to that of any conventional two wheel scooter of similar engine size. And in addition, of course, its weather protection (even though it had no heater) and load capacity was unmatched by any two wheeler, while the fuel consumption could be impressive. *Motor Cycling* reported that 'An overall figure of 90mpg was recorded, even when the model was cruised at near maximum speed on the open road for long periods'.

Two criticisms were reported – one of the level of noise and vibration experienced within the vehicle, and the other that the experienced motorcyclist might experience some difficulty in mastering the *left* hand twistgrip throttle control. Even so, *Motor Cycling* summed the machine up favourably in these terms: 'Unusual in layout, revolutionary, perhaps, in conception, the Messerschmitt 'Kabinenroller' proved itself to be a thoroughly practical vehicle, possessed of a degree of economy and performance far above average'.

By the time Germany's second post-war international motorcycle show took place at Frankfurt that October, sales of the KR175 Messerschmitt cabin scooter were reported to be enjoying a considerable degree of success. April 1954 witnessed the first to arrive in Britain with the appointment of Beulah Hill Engineering of London SE19 as concessionaires for the marque. And by mid-1954 some 200 KR175s were being produced every week at the Regensburg works. The British price by the end of the year was £325 18s 2d, with optional extras including a spare wheel, chromed hub caps and a luggage rack.

10,666 KR175s had been made by January 1955 when the model was replaced by a new version, the KR200. The larger engine capacity of 191cc was achieved by increasing the bore to 65mm, and with the same compression ratio, power was increased to 9.7hp at 5250rpm. This took the maximum speed up to a claimed 65.5mph although *Motor Cycling* only managed 56mph on a level road on full throttle. But a further improvement was the inclusion of a reverse gear and independent suspension on all three wheels by a combination of rubber springs and hydraulic shock absorbers. At the rear, the power unit, together with the gearbox and an 11 litre fuel tank, were mounted with the wheel on a tubular steel trailing arm pivoted on bearings either side of the vehicle, a rear cowling hinged upwards to give access. Meanwhile the reverse gear was actually *four* reverse gears selected by the rather eccentric method of restarting the engine via a second ignition system so that it ran backwards, with all the forward ratios operative.

More attention was paid to comfort by fitting a heater – and at the same time, the rather cramped rear accommodation of the earlier model was revised, a wider seat supposedly giving space for both an adult and a child in the back. Also a little eccentric was the fitting of both the heater control and the petrol tap by the rear seat.

1955 was the year when West Germany was granted sovereignty and permitted to re-arm, allowing Messerschmitt to return to aircraft production. Fend then started

And this is what made it go – the Fitchel & Sachs fan-cooled two-stroke single, with Siba electric starter

to look for a way whereby he could take control of the Regensburg production, and with the aid of the Bavarian State Bank eventually formed Fahrzeug und Maschinenbau GmbH Regensburg (FMR) and gained the right to continue manufacture using the Messerschmitt name.

DOUBLE
BUBBLE

In this second part of the Messerschmitt story,
Mick Walker recounts the rosy years.

*Very collectable; the Sports Cabriolet
complete with mock lizard trim and chrome
'bullet' holes in engine compartment.*

1955 was to prove the peak year for Messerschmitt's unique Kabinen-roller in both production and publicity terms. With production at record levels, the company built a very special machine called the 200 Super based on the standard KR200 2½ seater, with the objective of testing the speed potential of the design.

The Super had aircraft fuselage lines produced simply by fitting a single, double-curvature fairing in place of the tourer's cabin top. This carried a moulded windscreen at the front and formed a streamlined headrest for the driver at the rear. The only other external modification was an alteration to the shape of the front wings in order to improve their drag factor, but otherwise the bodywork, steering, suspension and dampers of the machine were claimed to be completely standard.

Modifications in the engine room had raised the power of the 191cc fan-cooled engine to a maximum of 14hp at 5500rpm using normal pump petrol in a 25 to 1 petroil mix. The power boost had been achieved by raising the compression ratio

from the standard 6.6 to slightly over 8 to 1, as well as enlarging and polishing the inlet and exhaust ports. Although (perhaps rather surprisingly) the standard silencer was retained, a 27mm Bing carburettor was fitted in place of the 24mm instrument with air filter used by the KR200.

In a highly successful sortie at Hockenheim on 29/30 August 1955, this machine set three new world records in the under-250cc three-wheel class, which stood in the 350 class as well! The records were for the 1000 miles (at 65.754mph), 2000 kilometres (at 65.24mph) and 24 hours (at 65.03mph). Weather conditions on the day on the record run were not ideal, nor was the machine extended unduly, for the initial aim had been to test steering and suspension at high speed with the records as a by-product. The real performance potential of the *Rekordwagen* was indicated better by its practice results which included a lap average of 72.5mph over the 4.8 mile circuit and a flat-out top speed of exactly 80mph.

Messerschmitt's own account of the record run at the time gave an impression of

faultless organisation. After the intensive practice results had been studied, the target average speed for the record was set at 65.625mph. The team of six drivers were forbidden to lap at more than 68mph, but in fact the average at the end of each hour for the first 12 hours fell to 64mph because thick ground mist forced drivers to slow during the night run.

Despite this success, by mid-1956 Messerschmitt were suffering a sharp decline in demand – like many other German makers of three (and two wheelers. However, the Regensburg concern felt this far less than the majority of other marques. Not only had the KR200 sold steadily, but in 1955 they had acquired the German rights to the Italian Vespa scooter, formerly held by the now-defunct Hofmann company. Production of the Vespa under licence by a separate company Messerschmitt-Vespa GmbH at Augsburg greatly added to the firm's ability to survive where others were forced out of business.

Two versions of the Vespa were manufactured by Messerschmitt from 1955 to 1958, both using the Piaggio-designed

145.5cc (57 x 57mm) single cylinder two-stroke unit. The standard model was code-named Me55 and had a power output of 5.5bhp giving a maximum speed of 53.1mph with a dry weight of 94kg and a 2.6 litre fuel tank. The more sporting Grand Sport had an 8hp engine and weighed 102kg. Its top speed was 59.4mph. Both models had the same basic chassis with the engine mounted on the right. And unlike the earlier Hofmann, they had the headlamp mounted on the handlebars rather than the mudguard.

Another fillip for the company came in 1956 with another co-operative Italian-German venture. This time it was Messerschmitt's design which was to be built under licence in Italy. In the three wheeler sphere, nothing could be seen as more Germanic than the Kabinenroller, so it was rather surprising to find that the Messerschmitt design had been taken aboard by an Italian company – Mival, of Valtrompia, Brescia, well-known for their off-road motorcycles.

The Mivalino, as the design was called, was built entirely at Valtrompia and used the company's own 171.7cc two-stroke engine incorporating a reverse gear and fitted with 12 Volt electrics including a starter. This gave it a performance of 54mph, which was better than the original German 175, but no match for the KR200. It exhibited much original thinking and incorporated a number of differences from the German original, including an internal layout designed to suit local tastes in seating, dashboard layout and decor. A noticeable external difference was that the clear roof of the Messerschmitt had been discarded in favour of a painted top. The obvious reason for this was the protection from the burning rays of the Italian sun, which would have fairly roasted to a frazzle anyone who went motoring under a square yard of plastic 'greenhouse' in the height of an Italian summer!

The 1956 Frankfurt Show held in October saw Messerschmitt introduce an open touring version known as the KR201. The newcomer had a stubby windscreen and had an optional hood or tonneau – while the show model was lined out in imitation alligator skin!

In November, two British enthusiasts made a marathon journey across Europe and back by Messerschmitt KR200, to cover 2452 miles in six days and prove their and the machine's ability to cope with tortuous roads, poor fuel and arduous driving conditions. Sponsored by the new British importers, Cabin Scooters Ltd of London W1 and the dealer MPHW Sales, Michael Morris and Tony Jarman left London on the morning of Saturday 13 November, their eventual destination on the outward trip being Rome. The only mechanical problem they encountered was in Italy and was traced to worn dynamo brushes, and after this was fixed, they arrived back in England on Monday 19 without further incident. At the end of their journey although weary and glad to be home, they were nonetheless pleased that their confidence in the tiny vehicle had been upheld.

1957's only change to the Messerschmitt three-wheeler range was the introduction of the KR201. As on the Frankfurt Show model, this featured the lizard skin plastic interior trim and a pair of chromium-rimmed ports on each side of the engine housing cover. by March 1958, three models of the popular Cabin Scooter were on sale in Britain – the KR200 Standard at £325 6s 4d, the KR200 de Luxe at £339 13s 6d (including a heater as standard) and the KR201 Drophead at the same price.

But in Germany at the Frankfurt Show, a brand-new model appeared. This was a four wheel version powered by a twin-cylinder 493cc (67 x 70mm) Fichtel and Sachs two-stroke engine rated at 19.5hp at 5000rpm. This had larger 4.40 x 10in tyres and a much higher weight of 390kg dry. The TG500 Tiger, as it was called, offered a maximum speed of 78.1mph – it was widely

The Messerschmitt Owners Club is active at various classic shows up and down the country.

advertised as being able to do 90mph! – but the model was not a success and only some 320 examples were produced despite its being listed until 1961. Few were offered on the export market (eighteen were sold in Britain), but in an attempt to gain more power, one brave soul in Britain later fitted his KR200 with an engine from a 692cc Royal Enfield Constellation, providing a genuine 100mph – at least in a straight line!

October 1958 brought the news that 1959 would see the introduction of a new model, the Sports Cabriolet KR200 de Luxe, best described as a compromise between the soft top KR201 and the KR200 de Luxe. It retained the permanently-fitted sidescreens and sliding windows of the KR200, but the domed portion was replaced with a quickly-detachable and rapidly-fitted folding top. This could be fully-adjusted in several positions between fully closed and completely open. A modification to all models that year was that the rubber channels for the sliding windows were lined with velvet to prevent them sticking and the vertical edges were sealed with a weatherproof strip.

In 1959 the KR200 Standard was discontinued in Britain and the new Sports Cabriolet KR200 sold for £346 3s 0d. From that period on, the model range continued virtually unchanged until 1964. By this time Fahrzeug and Maschinenbau GmbH (Messerschmitt) had begun to be actively involved in the aircraft industry once again, and therefore no further development work was carried out to update the *Kabinenroller* concept – and in any case year by year demand was falling for the vehicle.

Fritz Fend left the company in November 1963 to work as an independent consultant with firms such as Fichtel and Sachs, but it is a fitting tribute to his unique design that not only was it the first German mass-produced three wheeler to be built, but it also outlasted all of its competitors. A total of 41,190 of the various versions of the basic KR200 were made from 1955 to June 1964 – and survivors are today considered as true classics in their own right.

Two British Enthusiasts made a marathon trip to Rome and back in 1956 covering 2452 miles in six days.

NORTH AMERICAN MUSTANG P-51D

It's the fastest Mustang we've ever tested, but is it a match for the dreaded Messerschmitt?

BY PETER EGAN
PHOTOS BY RICHARD M. BARON

PHOTO BY PHILIP MAKANNA

Lt. RICK BARON and I had just returned from the USO club and a game of darts with some of the Limey flyboys when Reynolds showed up in the wing commander's "borrowed" Jeep with a bottle of scotch in one pocket of his raincoat and a tinned fruitcake in the other.

"No boiled brussels sprouts tonight, boys," he said. "We can trade this hooch for some real eats." Reynolds is seeing an English nurse in the WAAFs whose old man is some kind of bigwig in procurement, with a title that goes back to King Arthur and a house that makes the Pickford and Fairbanks place look like it's strictly from hunger. He can get us everything but home.

We went into the ready room, where the Old Man, Col. Bryant, was dressing down one of the new guys for having used a dangling participle on the radio over Düsseldorf. Seems he also messed up "who" and "whom." Big trouble. It's not easy flying in a motorjournalist squadron, and the new kid may not make it. I heard him say "orientate" out on the strafing range.

We were about to slink off toward the officer's club bar when the Old Man called us into his office for a briefing. "Right now," he growled, in his usual genial manner.

"Here's the story, gentlemen," he said, leaning on his desk and squinting at us through a cloud of cigar smoke. "The top brass has sent down word there's a new Mustang on the way, a P-51D. It's got everything; speed, range, drop tanks, six Brownings. You three have been assigned to go stateside and look at the new bird. Some of the other units have had it since 1944. I want to know if we need it."

We all looked at one another. Art Officer Baron raised an eyebrow. "Seems to me, Sir, we're doing okay with the Mustangs we've got now," he said. "We've got the GTs with the HO 302 and the handling package. What more do we need?"

Baron has 21 kills, having just smoked a 325i last Monday in a roundabout at the Biggin Hill exit. He likes the GT.

"There's nothing wrong with the basic Mustang GT," Col. Bryant said, "but it's a car, for God's sake, and the P-51D is an airplane. Our B-17s are getting shot out of the sky because we can't give them any air cover. Our current Mustangs can drive as far as the cliffs of Dover and then they have to turn around. We want a Mustang that can go to Berlin and back. So pack your gear, gentlemen. You're leaving tonight for Mojave, California."

As we stood up, he said, "One more thing. The intelligence boys have reconstructed a crashed Messerschmitt and we're sending it along for comparison. Fly the P-51, look at the Messerschmitt and tell me what you think. I want to know if we have a chance against this thing."

We flew stateside in a homebound Liberator, did the usual buzz under the Golden Gate and found ourselves in Southern California. The guys dropped me off at Chino Airport, where the P-51 is kept, and then headed for Mojave Airport, up in the high desert, with our test equipment in the back of a Corvette ZR-1. Intelligence thought this was the only car

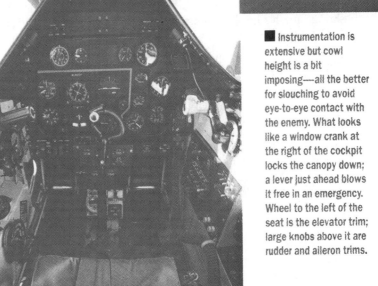

that had a chance of keeping up with the P-51 on its takeoff roll, to measure acceleration. (As usual, Intelligence didn't have quite enough.)

I was to meet the Mustang owner, Elmer Ward, at his hangar in Chino and fly up to Mojave with him. There we would rendezvous and conduct our secret comparison test with the Messerschmitt.

Elmer Ward drove up and we shook hands. Ward is a slim, energetic industrialist in his 60s, a Cal-Tech-trained engineer and owner of his own manufacturing firm. He also owns a company that builds new Mustangs from a vast supply of surplus parts he keeps at Chino. He

■ Instrumentation is extensive but cowl height is a bit imposing—all the better for slouching to avoid eye-to-eye contact with the enemy. What looks like a window crank at the right of the cockpit locks the canopy down; a lever just ahead blows it free in an emergency. Wheel to the left of the seat is the elevator trim; large knobs above it are rudder and aileron trims.

"Feats of derring-do like dragging my wingtip on the desert floor at 300 miles per hour will strike terror into the hearts of Luftwaffe pilots," thought our man, grinning impishly.

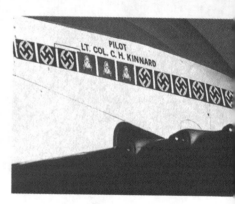

rolled open the hangar doors and showed me his own Mustang, a P-51D built in 1945, a real beauty.

As we looked the plane over, Ward gave me a little history. This particular P-51 was built in Inglewood, California, flown to England during the last months of World War II, and lent much later to Universal Studios for the 1956 movie *Battle Hymn*, with Rock Hudson. A man named Ascher Ward (no relation) bought the Mustang off the back lot in 1970 for $3500 and restored it to flyable condition. Elmer Ward bought it in 1975 and did a complete restoration in 1980.

The airplane is now painted in the colors of Col. Claiborne H. Kinnard, the famous ace (27 kills) and commanding officer of the 4th Fighter Group of the 334th squadron, 8th

Air Force in England. Kinnard survived the war and died at his home in Franklin, Tennessee in 1967. Ward painted the P-51 in Kinnard's colors out of admiration for Kinnard himself and for the 4th Fighter Group, which began life as the three original Eagle Squadrons, formed of American volunteers who fought for England before the U.S. entered the war. Ward actually flew the Mustang to Cleveland in 1976 to have Kinnard's *Man O' War* script painted on the nose by Don Allen, the original artist.

Enough history. Time to go flying and take some road test notes for the Old Man.

Ease of entry? Fair. You step on the tire and landing gear strut, climb up on the front of the wing, crank back the canopy and just step in, if you are the pilot. As a passenger, I

had to slither back under the canopy into a rear seat installed where the 85-gallon fuselage gas tank used to be. The plane still carries 180 gal. (usable) in the wing tanks, enough for about 800–900 miles of cruise. Under the wings and fuselage are attachment points for the auxiliary drop-tanks that extended the P-51's range deep into Germany, making life good for Allied bomber crews and bad for the Luftwaffe.

Parachute harnesses first, then seatbelt and shoulder straps. Elmer says if you bail out of this baby you pull a canopy release lever, duck low so the departing canopy doesn't take your head off, crouch on the edge of the cockpit and leap hard toward the right wingtip. This way, with luck, you will clear the tail so it doesn't break both your legs as it passes by.

A booster cable from an auxiliary power unit is hooked to the belly, Elmer calls "Clear!" and hits the starter button.

When the prop begins to turn, there's something startling about seeing those four huge blades jerk so quickly away from rest, the sheer mass and diameter of them—11 ft. 2 in. of aluminum alloy.

Two giant puffs of smoke explode from the six straight stacks on either side of the nose, and the supercharged 1649-cu.-in. V-12 (the displacement of 4.7 Corvette engines) is running and muttering like a heavy smoker getting up in the morning and clearing his lungs. Even at idle, there's an undertone of menace in the deep exhaust note. It says, "This is a warbird, pal, and engines don't sound this way unless they are going to war." You feel that the Beast has been awakened, and he's not happy about it.

We taxi between hangars, watching our wingtips and looking down on the roofs of cars. Five minutes' wait in the runup area while 21.2 gal. of 60-weight oil and 16.7 gal. of coolant warm up, and we taxi into position and hold. Throttle comes up, brakes are released, manifold pressure climbs to 50 in. (they used 61 in the war) and the Mustang thunders forward, trying to torque its tail sideways against Ward's steady rudder pressure. The tail comes up and we are flying.

Climb is so easy and lazy, it's almost an anticlimax; 200 mph, 2000 ft. per minute with the nose barely over the horizon. The great sink that contains Los Angeles drops from beneath us, and we are headed for the snow-dusted San Gabriel Mountains. A 240-mph (indicated) cruise at 2150 rpm takes us over the peaks at 9500 ft. The engine has the mellow, rhythmic shuffle of a sewing machine made from locomotive parts. The Mustang is over the high desert now, trimmed nose low for a wonderful view of ground and sky. Forward vision is excellent. The airplane has assumed its natural gait, which is somewhere between prowling and ranging. It cruises along head down, purring, but full of latent ferocity, no doubt wishing it had a train to strafe.

I lift the headphones experimentally away from my ears for a second and am assaulted by the loudest mechanical cacophony I've ever experienced. It's like . . . like what? Like having a stethoscope on a tin roof during a ball-bearing storm. The headphones go back on and I try to refocus my eyes. I won't do that again soon.

"Some aerobatics?" Elmer asks over the intercom.

"Sure!"

The brown/blue horizon tumbles and tilts. Eight-point rolls, 4-point rolls, barrel rolls, right and left knife-edge, Cuban 8 and a wonderful bank reversing maneuver called a John Derry Turn that I've never done before. Elmer Ward does them all with a velvet touch and confident precision that would reduce even my old aerobatics instructor to an appreciative silence. Twisting and turning over the desert floor, the Mustang has the lazy, muscular confidence of a Great White shark, casually deadly and in no particular hurry. It feels stable, powerful and solid.

We bank off toward Mojave airport, not far from where the space shuttle lands at Edwards, spiral down and swing in for a fast but gentle 3-point with all 12 pipes crackling like a string of Black Cats on the 4th of July. We taxi, canopy cranked back, breathing clouds of heady rich 100-octane avgas fumes. Elmer parks and shuts her down. The sudden silence is deafening.

"This is a wonderful airplane," I say, stating the obvious.

Ward agrees. "I think it's the best fighter of WWII. It's strong and simple—just four big longerons with skin riveted to them and a big engine in the front. It was also the cheapest fighter of WWII, $50,000 apiece, without engine."

We climb down from the wing and Reynolds is there, with his Corvette and test gear. The Messerschmitt is arriving tonight, Reynolds tells us. We test in the morning.

At sunrise, our friend Paul Prince arrives with his two sons, Austin and Zeke. They have brought their 1953 Messerschmitt KR-175 over from Santa Barbara in an unmarked white rental van. Smart. We unload the little ship, and, frankly, she doesn't look like much next to the Mustang, but you never know. The Messerschmitt is powered by a 9-bhp Fichtel & Sachs 2-stroke single, the Mustang by a 1400-bhp Rolls-Royce Merlin V-12. Is this the Messerschmitt that Col. Bryant had in mind? It doesn't even have wings.

Testing is somewhat inconclusive. While the Messer is more agile through the cones, the P-51 simply blows them away with prop blast. Which technique is better? I'll take the P-51.

Lateral acceleration? We didn't have access to a skidpad at Mojave, but the P-51 generates the usual 4 to 5 g's in simple aerobatic maneuvers and is structurally rated far beyond the ability of humans to withstand g-forces and remain conscious. I doubt the 4.00 x 8-in. tires on the KR-175 can equal this feat. The Messerschmitt produces no blackout, even in hard turns. The KR-175 is also slow, with a top speed of about 48 mph, versus the Mustang's 428 mph.

About all the Messerschmitt really has going for it is good fuel mileage (102 mpg vs. 5.0 mpg for the P-51),

■ Fear of being sucked into the P-51's prop motivated our man to post what is possibly the fastest slalom time ever achieved by a car powered by a Fichtel & Sachs 2-stroke.

NORTH AMERICAN MUSTANG P-51D

Takeoff roll	**1500 ft**
0–40,000 ft	**25 min**
Top speed	**428 mph**
Dive speed	**505 mph**
Vertical accel	**8.0g**
Landing roll	**1800 ft**

PRICE

List price,
FOB Inglewood, Calif ... **(1945) $50,985**   Price as tested .... **est $75,985**
Price as tested includes std equip. (bulletproof windscreen, ejectable canopy, six Browning 50-caliber machine guns, drop tanks, cruise control), Packard-Merlin V-1650-7 engine (est $25,000).

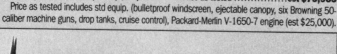

ENGINE

Type	two-stage supercharging, inter & after coolers, alloy block & head, **V-12**
Valvetrain	sohc 4-valve/cyl
Displacement	1649 cu in./ 27,022 cc
Bore x stroke	5.40 x 6.00 in./ 137.2 x 152.4 mm
Compression ratio	6.0:1
Horsepower (SAE):	**1400 bhp @ 3000 rpm**
	1720 bhp @ 3000 rpm
	(during war emergency; 5 min. max)
Bhp/liter	51.8
Maximum engine speed	3240 rpm
Fuel injection	Bendix PD-18C1
Fuel	premium leaded, 100 pump oct

DRIVETRAIN

Transmission	**constant speed propeller**
Type	Hamilton Standard 4 blade
Diameter	11.2 ft
Pitch	variable, 23–65 deg
Actuation	automatic, hydraulic
Cruising speed @ rpm:	310 mph @ 2300 rpm
Maximum speed (level flight)	428 mph @ 17,000 ft
Maximum speed (dive)	505 mph

CHASSIS & BODY

Layout	**front engine/front drive**
Fuselage & wings	aluminum skin on four longerons, main & rear wing spars, 25,000 rivets
Brakes, f/r	**7.5-in. discs/none,** flap assist
Wheels	cast alloy; **14 in. f, 5.5 in. r**
Tires	Uniroyal Aircraft, **27 x 10 in. f;** McCreary, **12.5 x 4 in. r**
Steering type	**(ground) tailwheel, (air) 10.4-sq-ft rudder**
Suspension, f/r:	retractable; **pneumatic struts,** oil damping

HANDLING & BRAKING

Vertical accel	8.0g positive
Balance	affected for weeks
Speed thru 700-ft slalom	unknown, prop blows all the cones over
Minimum stopping distance From 87 mph (landing speed)	1800 ft
Overall brake rating	nothing broke

Subjective ratings consist of black and white with no shades of gray. This is World War II, pal!

GENERAL DATA

Curb weight	**8320 lb**
Test weight	**8520 lb**
Weight dist, f/r, %	**92/8**
Wheelbase	16.3 ft
Track, f/r	11.8 ft/0 ft
Length	32.2 ft
Width	37.0 ft
Height	8.7 ft
Trunk space	7.9 cu ft per wing or six machine guns & 2080 rounds

INSTRUMENTATION

700-mph speedometer, 4500-rpm tach, oil press., oil temp, fuel press., coolant temp, induction air temp, +8/−4g vertical accelerometer, manifold press., dual RMI, suction press. for artificial horizon, artificial horizon, 40,000-ft altimeter, turn & slip indicator, rate of climb indicator, horizontal situation indicator, oxygen system press., compass, dual fuel level indicators

FUEL ECONOMY

Normal flying (300 mph)	5.0 mpg
EPA rating	4.4 city recon/ 4.4 highway strafing
Cruise range	900 miles
Fuel capacity	184 gal.
Oil/filter change	125 hrs
Tuneup	100 hrs
Basic warranty:	unlimited missions or VE Day

ACCELERATION

Runway time to speed
0–30 mph	Corvette ZR-1-like
0–60 mph	the Vette is doorless
0–80 mph	no contest
0–100 mph	almost airborne
Rotation	1500 ft @ 110 mph

Time to altitude	Minutes
0–10,000 ft	3.7
0–20,000 ft	7.0
0–30,000 ft	12.6
0–40,000 ft	25.0
Service ceiling	41,900 ft

INTERIOR NOISE

Idle	95 dBA
Maximum, takeoff	125 dBA
Constant 240 mph	115 dBA

easy entry and parking, and low maintenance costs (a Merlin V-12 rebuild costs $45,000; the Fichtel & Sachs 175 could probably be rebuilt for around $100).

Measuring 0–60 acceleration on both turned out to be a bust. The Messerschmitt won't go 60, and our Corvette couldn't keep up with the Mustang. When we took off to leave Mojave, the Corvette ZR-1 and its 5th wheel lined up at the P-51's right wingtip, hoping to pace the Mustang and measure its performance. The smart money said the Corvette would be loafing until 60 or 70 mph. The smart money was wrong.

Elmer Ward ran the Merlin up to 3000 rpm and 30 inches of manifold pressure, released the brakes and shot down the runway, building to 50 in. of pressure. The Corvette never had a chance. It stayed with us until about 40 mph and then disappeared into the Mustang's twin mirrors like something tied to a post. We swooped into a climbing 180-degree turn at 200 mph and seconds later were flying knife-edge, looking down our wingtip at the Corvette's roof. The Messerschmitt stood by the side of the runway, growing smaller and smaller in the vastness of the desert until it was lost among the rocks and tumbleweeds.

It's easy to see now why the P-51 has been such an overwhelming success in the other squadrons. I think the Old Man is going to like this new Mustang. I think he's going to like it a lot.

The Messerschmitt boys are going to wish they'd voted for Roosevelt. 🏁

Test Notes . . .

■ Absolutely the most powerful Mustang we've ever tested, the P-51D's prodigious 1400 bhp produces nary a chirp from its tiny tires off the line. Would likely better 110 mph if not for wings mounted upside down causing it to fly.

■ Absolutely the loudest Mustang we've ever tested, the P-51D's going to collect plenty of fix-it tickets in hospital zones. Speaking of micro-surgery, the Mustang's radiator fan can make for a nasty nick. Ouch!

CONTINUED FROM PAGE 149

records at Germany's Hockenheim race circuit.

Maybe it was this success that spurred the factory to consider a more powerful version. At the 1957 Frankfurt Show the TG500 Tiger was unveiled — a four-wheeled version with a twin-cylinder 500cc Sachs-designed two-stroke.

This had 10-inch wheels instead of the KR200's 8-inch rims. It was advertised as being capable of 90mph, but Tiger owner Mark Smith of the Messerschmitt Owners' Club reckons this to be rather optimistic.

"They are capable of over 70mph," said the Southern chairman at the club's rally last year. "We think 300 were made in total. Eighteen were sold in this country and we can trace them all. They are so rare it's not realistic to talk about current values, but one was advertised recently for over £10,000."

Tricycle undercarriage

Before taking a closer look at the KR200 let's briefly trace its ancestry.

Enter German engineer Fritz Fend (b1920). During the war his noteworthy skills were called upon by the German air force.

He was particularly involved in designing tricycle undercarriages for Messerschmitt fighters and bombers and so came into contact with Professor Willy Messerschmitt and other eminent engineers.

At the end of the war Germany was forced to cease munitions manufacture. Fend returned to his home town of Rosenheim, Bavaria, and set up a technical consultancy.

He designed a three-wheeled hand-propelled invalid carriage for wounded who had lost both legs.

Next he added a 38cc two-stroke motor, and then built more refined open and covered invalid carriages with 100cc motors, plus a freight scooter.

Then, in 1951, a chance meeting with Professor Messerschmitt led to an arrangement to make Fend three-wheelers at Messerschmitt's Regensburg aircraft plant. Work on the KR175 began in 1952 and the first version was ready the following year.

In 1955 West Germany regained sovereignty and was allowed to re-arm. Professor Messerschmitt went back to aircraft production while Fend sought financial help from the Bavarian State Bank to take control at Regensburg.

His negotiations were ultimately successful and he formed FMR (Fahrzeng und Maschinenbau GmbH Regensburg) with fellow engineer Valentin Knott and secured the right to continue using the Messerschmitt name.

But growing affluence in Germany meant the Messerschmitt's card was marked. People wanted larger and more comfortable vehicles.

The last Tiger was produced in 1962, and the last KR200 in June 1964. Fend had resigned from FMR in November 1963 to set up an engineering consultancy. Clients included Fichtel and Sachs with work concentrating on the Wankel engine.

Now let's take a closer look at the anatomy of a KR200 Messerschmitt.

Chassis and body

The chassis is of tubular steel to which is mounted the pressed-steel body. The cockpit is hinged and features a perspex dome with sliding windows for ventilation. The front area has a safety-glass panel with wiper.

The body was available in four versions: Hardtop, Roadster, Cabrio and Sport.

The Cabrio, introduced in 1959, is a converted Hardtop. The Roadster features cutdown sides allowing the driver to rest his arm on the door top as he motors along. It has a fixed front screen and a hood which can be pulled over to fasten to the screen.

The Sport is the only version without door. It has a flyscreen and tonneau cover. For outdoor types only, you could say.

Steering — via a handlebar — is direct and consequently is reported to feel rather heavy until the driver gets used to it.

The centrally located steering column mates with a divided track rod via a lever welded to the column at floorpan level. Turning circle is 27ft 2in. All three wheels are independently sprung on soft rubber torsion springs with hydraulic shock absorbers. The engine does not move with the rear wheel's trailing arm.

The KR200's predecessor, the KR175, is without suspension but does have a sprung driver's seat. Owner of one of the few 175 examples in the UK, Nicholas Oddy from Edinburgh, cracked the dome with his head when crossing the grass field at the 1986 rally!

All three wheels have cable-operated drum brakes of $4\frac{1}{2}$in diameter. The handbrake, operated by the driver's left hand, also operates on all three wheels.

The engine is a Sachs fan-cooled single-cylinder two-stroke with a cast-iron barrel, alloy head, 65x58mm bore and stroke, 6.6:1 compression and 9.7bhp claimed at 5250rpm. There is a Bing carburettor with intake silencer and filter. The engine runs on 24:1 petrol/oil mix from a three-gallon fuel tank.

Transmission

Four forward gears and four reverse! Yes, four reverse. Selection is by a lever which the driver operates with his right hand. The driver moves the lever forward to change up and the gearbox features a motorcycle-type positive-stop mechanism.

Foot controls are the same as in a modern car. The left pedal operates a four-disc clutch.

The gearbox is in unit construction with the engine. First to fourth ratios are: 4.2, 6.1, 9.1 and 17.7. Final drive is by enclosed chain in an oil bath. There is a 12-volt lighting system with 35/35 watt headlights, and ignition by coil, contact breaker and battery. Starting and battery charge is provided by a 135-watt Siba Dynastart unit. Incorporated in this is a second ignition system and switching facility so that the engine can be started to spin in reverse and so give four reverse gears.

Interior

Two's company and three's a crowd under the dome. The driver's seat is quite narrow so that the rear passenger has space to stretch his or her legs either side. The wider rear seat is said to be able to accommodate an adult and child.

The petrol tap with reserve position is situated by the rear seat. So, too, is the heater control knob. This you pull out to allow engine heat to reach the driver via a flexible hose.

Dearly loved

So there you have it — in brief, you might say. The Messerschmitt is dearly loved by 400 or so members of the Messerschmitt Owners' Club in the UK.

Members are grateful to joint-president Les Tilbury from Frating Green, Essex, whose spares service has kept so many on the road.

Phil Boothroyd, the other joint-president, reckons there are about 600 Messerschmitts in Britain. Phil is researching a book on the marque and provided much of the information for this feature. If you can help with useful information, drop him a line at 45 Welford Road, Shirley, Solihull, West Midlands.

Minimal frontal area provides respectable performance from the 10bhp Sachs engine – and excellent fuel consumption

Our thanks to the three Messerschmitt stalwarts who turned up for our photo session: Phil Boothroyd, Ron Crawley and Les Tilbury; and to Tony Ditheridge for allowing us the use of his Pilatus P2 Me109 lookalike. Thanks also to Ipswich airport for allowing us to park the *Jagdstaffel* in front of their terminal building!

Form vs Function

Is this a case where Wright is wrong?

BY NORM MORT, WITH PHOTOGRAPHY BY STUART BEATTY

ACCORDING TO ARCHITECT Frank Lloyd Wright, "Form follows function," but when it comes to these automobiles it's a matter of form versus function.

There are a substantial number of enthusiasts who, at first glance, have an almost polarized taste in vehicles. Peter Svilans' love for the shapely "hairy chested brute", the big Healey, and his very functional micro-sized Messerschmitt must surely fit into this category.

These very different vehicles attempt to answer the same question, "How do you transport two people from point A to point B?" in two very different ways.

It's been almost forty years since Donald Healey's new sports car was first shown to the public at the annual Earls Court Motor Show in 1952, and yet the shape

remains as fresh and stylish as ever. As a functional vehicle it has many shortcomings. The front bucket seats are not the most comfortable over the shortest of distances. The two so-called occasional seats are at best a joke, but I suppose no less crowded than those found in our coupes today. (So much for progress). The shift

Bubble canopy gave excellent visibility, even when open! (above) Peter Svilans with his "Odd Couple"

is different from the Healey 3000 in that it protrudes from one side of the transmission. Long and spindly, it is quite precise and smooth. The large diameter steering wheel would make most of the "not yet thirty-something" enthusiasts shudder. The handling and ride are truly "classic vintage"; slight understeer going into a turn and changing to oversteer with more throttle.

The majority of readers are well aware of the great Donald Healey, his company, and his cars, but the Messerschmitt, and those involved, have never attained similar status.

Willy Messerschmitt began by building sailplanes and sports planes in 1923, and by the end of WWII was employing 45,000 workers. With many of his factories destroyed and the ban on aircraft manufacture at the end of the war, Messerschmitt maintained the remnants of his factories by producing sewing machines and auto parts. Then, in 1952, aeronautical engineer Fritz Fend approached Messerschmitt with the idea of building the Fend Lastenroller (cargo carrier) in his Regensburg works. At the same time Fend carried on developing a model based on a moderately successful, single seat, "Fend Flitzer" for which he needed capital and facilities to produce. From this the Messerschmitt Kabinenroller (cabin scooter) was born. It may have initially appeared rather unsophisticated with its rope-pull start, manual windscreen wiper, handle bar shift, and lack of reverse. (These small inconveniences disappeared on the KR 200), but there was an ingenious side to the design. The chassis was of a tubular layout that all Formula car manufacturers later accepted universally. The unique, lightweight design provided adequate performance and comfort along with 60 mpg economy.

The KR 175 was replaced by the more powerful and convenient KR 200 in 1955, and that year proved to be the company's most successful; building just under 12,000 units. Then in 1956 the ban on German aircraft construction was lifted and Willy Messerschmitt lost interest in the KR. Fend was now in a better financial position and purchased the firm and the works. New models were introduced under the FMR nameplate. A Kabrio, a roadster, a sports model and eventually a 4-wheeler; the Tiger Tg500.

These post-war, basic transportation vehicles were a boon to poverty stricken

Germany. All-told 50,000 were produced by 700 employees before the Volkswagen and Mini finally hit the used car market, and demand dwindled to nothing in 1962.

Getting into an FMR can be an adventure in itself. It appears to be fragile and one is almost afraid to put all ones weight in or on that little car. Both feet in first, then sit and the canopy is dropped. You are surprised, (not your first or your last in this case!) by the amount of space left, and it's not claustrophobic! Going from forward gears to reverse gear(s) necessitates switching off the ignition. After the initial shock (a step up from surprise) it all makes quite good sense. Besides, the Dynastart (combination starter and generator) works well. The engine doesn't sound as "bad" as you might think. It's not like a sewing machine or your lawnmower – well maybe, if they were supertuned! If you analyze acceleration, g-forces, aerodynamics, etc., then you're missing the point. Heck, how can you think about that kind of stuff when you're squealing with delight, laughing uncontrollably and waving to every Tom, Dick and.. wow, hello Sally!

The Odd Couple

The diminutive Messerschmitt is about as minimal a vehicle as one could find, yet still be able to venture out into the real world. At the same time it is able to fulfil the basic transportation needs of its owner with a certain amount of convoluted style. This is the other side to the Messerschmitt's attraction, for beyond its practicality the overall design has the charm of an Annie Hall. In contrast the

Healey is a car built to appeal to the senses all of them! Depending on your preference it's, "the naughty girl who lives on the other side of the tracks, or, the rocker in the black leather jacket who rides a motorcycle". There is a strong sensual attraction to the rugged beauty of its soul.

This automotive odd couple's sex appeal is something Frank Lloyd Wright would have had to speak to Sigmund Freud about. On the other hand, auto enthusiasts have no problems accepting their earthiness, particularly if they have an artistic side, as is the case with owner Peter Svilans.

Svilans' early years were spent in England and Germany but before reaching his adolescent years his family moved to Milton, Ontario. By sixteen, Svilans was a serious auto enthusiast who had already restored a '59 Mini. A 1966 MG Midget, an Austin 1100, and a Fiat 850 Coupe followed, but once enrolled in the Ontario College of Art his automotive interests began to change. "As a student in industrial design I began to look at form as well as function - form penetrating space". He saw the Austin Healey and the Barchetta Ferrari as, "the epitome of styling - a high point!"

He recalled, at the age of ten, of being "bowled over" by the style of a yellow Austin Healey 100-6. The shape has, "a tension to it". There are no frivolous excesses, no add-on styling cues, and no trickery of design.

The Messerschmitt was also introduced into Svilans' life early. Apart from seeing them in Germany, Peter remembered an enthusiast terrorizing the neighbourhood in Milton. He sees the Messerschmitt's functional design as being secondary to its overall visual and mechanical appeal. It was originally conceived as a source of cheap transportation for the post-war worker and labourer in war-torn Germany, but as a "new prosperity" began to evolve the Messerschmitt became, particularly in North America, a vehicle sold generally to eccentrics, artists, and photographers.

Peter noted, "A Florida enthusiast with a long white beard bought the remaining 'Schmitt dealership. If you need parts he has a shed deep in the swamp. His wife carries a pistol at all times to shoot the alligators".

As well as eccentrics the cars did appeal to creative types as even The King of Rock n' Roll had a 'Schmitt. Elvis is quoted to have said in an issue of Rod Builder and Custom, "This is a cool little buggy if ever there was one, perfect for zooming

Form vs. Function

Austin Healey 100-6 BN4, BN6 (2-seater)
(July, 1956 - March, 1959)

Built: 14,435
Engine: BMC C-series, In-line 6 cylinder, 2639cc ohv pushrod, twin SU carbs (Front engine/rear wheel drive)
BHP: 1956, 102 @ 4600 rpm
1957, 117 @ 4750 rpm
Transmission: 4-speed, plus overdrive
Brakes: Girling hydraulic
Wheels: 4J disc or optional wire, 5.90 X 15
Weight: 2440 lbs
Performance
Maximum Speed .... 103 mph* (105)
0-40 mph .... 6.1 seconds* (6.1)
0-60 mph .... 11.6 seconds* (12.2)
0-80 mph .... 21.3 seconds* (22.5)
1/4 mile .... 75 mph ......18.2 seconds* (18.2)
*Sports Cars Illustrated 2/57
( )Road&Track 1/57

Messerschmitt
(1953-1962)

Built: 50,000 (45% exported)
Engine: Fichtel and Sachs Two-stroke, single cylinder 191cc air-cooled
BHP: 9.7 @ 5000 rpm
Transmission: 4-speed, reverse electrically selected
Brakes: Mechanical (cable), hand brake on all 3 wheels
Wheels: 3 disc wheels, 4.40 X 8
Weight: 506 lbs
Performance
Maximum Speed: ....perhaps 60 mph
0-40 mph .... 25 seconds (KR-200)
1/4 mile .... 42 mph .... 29.7 seconds

around town when I'm home. And she gives me almost 50 miles to the gallon".

It was Peter's artistic side that attracted him to the Messerschmitt. "I appreciate vehicles for what they are, what they try to be, and how they do it. The Messerschmitt is not a Buick or an Aston Martin. You have to appreciate it for what it is". Fritz Fend was solely responsible for the Messerschmitt design, as was Donald Healey for his successful sports car. It was not designed by a committee and it was perhaps this fact, more than any other, that made these two extreme designs attractive.

Taste, style and appeal are always very personal when it comes to cars, but as Peter pointed out, "Unique cars are one person's design and reflect part of their character".

Svilans' Antipodean Designs

In 1972, when Peter Svilans was an art major at the University of Guelph he purchased, "a decrepit green, two-seater, 1958 BN6 100-6", from a sports car dealer outside of Hamilton. I was told this $200 car ran perfectly, but soon discovered the engine was seized and cracked, the frame was broken, and the trunk was pushed in. Nevertheless my interest was completely focused on this car, and Art School took on a different meaning".

He read all he could find on the 100-6 and, "...made a list consisting of door hinges, emblems, etc, needed, ...but in reality the car was a total wreck". If the

Healey were to be made structurally sound, it was to require restoration from the inside out. Over a decade the car was dismantled, sand blasted and re-built. Apart from the painting he did everything himself including the cutting, fitting, and stitching of the upholstery. The enamel paint has weathered well over the past ten years, as has the car, "I wouldn't hesitate to drive the car to California".

In fact, Peter drives the 100-6 every season to the Austin Healey North American Conclave. "Over the years we've travelled over 50,000 miles to Kentucky, Virginia, Wisconsin, etc.; often with our daughter Alana in a crib in the back. Even after working around Healeys for 18 years I think the 100-6 is the best shape. The design was never bettered as far as looks go."

In the 1970s Healeys were not the valuable collector cars they are today. Many enthusiasts were dropping in V-8s rather than re-building pristine originals. Peter's restoration was complete but not on par with the concours standards of the 1990s. "Because it was built as a road car, lead was used and parts were brazed". Eventually, when a complete restoration is deemed necessary, Peter will turn this shapely form to its original showroom condition.

Peter's Messerschmitt is, in reality, a 1960 FMR KR200 that was discovered in a farmer's field outside Chatham three years ago. "It was a rusted wreck! I began to research 'Schmitts, and joined the Micro and Mini Car Club in California, the Messerschmitt Deutchland Club in Germany, the British Messerschmitt Owners' Club and the Canadian-based Isetta/Messerschmitt Club.

The Messerschmitt is now one of the finest examples in North America. He has restored his FMR to very high standards, replacing all the body panels, using only hammer welding, and fitting a genuine, optional, imitation snakeskin interior. (Peter is Precision Sportscar Servicing's upholstery and convertible top expert).

The 191 cc, two-stroke, Sachs Fichtel engine was rebuilt as were all the other mechanical bits and pieces. The black convertible top was hand made by Peter, and compliments the contrasting red bodywork. The "Kabine" licence plate is German for, "cabin scooter", a term often used to describe the vehicle.

Who says basic, inexpensive, and functional transportation has to be technically boring, and shaped like a box! As for Frank Lloyd Wright; I can't recall even seeing a garage in one of his designs. ∎

(top left) Tiny 191cc engine tucked away at the rear, complete with four reverse speeds! (top right) Looking more like the cockpit of a WWII Messerschmitt fighter, this car is ready to fight the traffic. (middle) Side view shows surprisingly aggressive stance of little Messerschmitt. (bottom) Contrasting designs are very evident – Form vs Function indeed!